PSpice for Digital Signal Processing

PSpice for Digital Signal Processing
Paul Tobin

ISBN: 978-3-031-79766-8   paperback
ISBN: 978-3-031-79767-5   ebook

DOI 10.1007/978-3-031-79767-5

A Publication in the Springer series
SYNTHESIS LECTURES ON DIGITAL CIRCUITS AND SYSTEMS #11

Lecture #11
Series Editor: Mitchell A. Thornton, Southern Methodist University

Library of Congress Cataloging-in-Publication Data

Series ISSN: 1932-3166   print
Series ISSN: 1932-3174   electronic

First Edition
10 9 8 7 6 5 4 3 2 1

PSpice for Digital Signal Processing

Paul Tobin

School of Electronic and Communications Engineering
Dublin Institute of Technology
Ireland

SYNTHESIS LECTURES ON DIGITAL CIRCUITS AND SYSTEMS #11

ABSTRACT

PSpice for Digital Signal Processing is the last in a series of five books using Cadence Orcad PSpice version 10.5 and introduces a very novel approach to learning digital signal processing (DSP). DSP is traditionally taught using Matlab/Simulink software but has some inherent weaknesses for students particularly at the introductory level. The 'plug in variables and play' nature of these software packages can lure the student into thinking they possesses an understanding they don't actually have because these systems produce results quickly without revealing what is going on. However, it must be said that, for advanced level work Matlab/Simulink really excel. In this book we start by examining basic signals starting with sampled signals and dealing with the concept of digital frequency. The delay part, which is the heart of DSP, is explained and applied initially to simple FIR and IIR filters.

We examine linear time invariant systems starting with the difference equation and applying the z-transform to produce a range of filter type i.e. low-pass, high-pass and bandpass. The important concept of convolution is examined and here we demonstrate the usefulness of the 'log' command in Probe for giving the correct display to demonstrate the 'flip n slip' method. Digital oscillators, including quadrature carrier generation, are then examined. Several filter design methods are considered and include the bilinear transform, impulse invariant, and window techniques. Included also is a treatment of the raised-cosine family of filters.

A range of DSP applications are then considered and include the Hilbert transform, single sideband modulator using the Hilbert transform and quad oscillators, integrators and differentiators. Decimation and interpolation are simulated to demonstrate the usefulness of the multi-sampling environment. Decimation is also applied in a treatment on digital receivers. Lastly, we look at some musical applications for DSP such as reverberation/echo using real-world signals imported into PSpice using the program Wav2Ascii. The zero-forcing equalizer is dealt with in a simplistic manner and illustrates the effectiveness of equalizing signals in a receiver after transmission.

KEYWORDS

Digital frequency, sampling, delays, difference equations, z-transform, FIR and IIR filters, convolution, windowing, filter design methods, Hilbert transform, decimation and interpolation, digital receivers.

I would like to dedicate this book to my wife and friend, Marie and sons Lee, Roy, Scott and Keith and my parents (Eddie and Roseanne), sisters, Sylvia, Madeleine, Jean, and brother, Ted.

Contents

1. **Introduction to Digital Signal Processing** 1
   1.1 Sampling 1
   1.2 Aliasing 1
       1.2.1 The ADC Parameters 2
       1.2.2 Digital Frequency 3
       1.2.3 Digital Samples 4
   1.3 Delay Unit and Hierarchical Blocks 5
       1.3.1 Transmission Line Delay 7
       1.3.2 Laplace Delay 8
   1.4 DSP Signals: Unit Step Signal 9
       1.4.1 Unit Impulse Signal 9
       1.4.2 Decaying Exponential Signal 11
   1.5 Digital Sequences 12
   1.6 The z-Transform 14
       1.6.1 Unit Impulse 14
       1.6.2 Unit Step 14
       1.6.3 Unit Delay 15
   1.7 Exercises 15

2. **Difference Equations and the z-Transform** 17
   2.1 Linear Time-Invariant Systems 17
   2.2 Difference Equations 17
   2.3 Digital Filter Classification 18
   2.4 First-Order Fir Filter 18
       2.4.1 z-Transforms and Difference Equations 20
   2.5 Pole-Zero Constellation and Bibo Stability 21
   2.6 Cut-Off Frequency 22
   2.7 Infinite Impulse Response Filter 24
       2.7.1 First-Order Low-Pass IIR Filter 24
       2.7.2 Cut-Off Frequency for the IIR Filter 25

2.7.3   IIR Phase Response 26
2.8   High-Pass IIR Filter 26
2.8.1   The −3 dB Cut-Off Frequency 27
2.8.2   Passband Gain 27
2.9   Step Response 28
2.10   Impulse Testing 29
2.11   Bandpass IIR Digital Filter 31
2.11.1   Bandpass Pole–Zero Plot 31
2.11.2   The −3 dB Cut-Off Frequency 32
2.12   Bandpass Impulse Response 35
2.13   Simulating Digital Filters Using a Netlist 35
2.14   Digital Filters Using a Laplace Part 37
2.14.1   Third-Order Elliptical Filter 38
2.14.2   Group Delay 40
2.15   Exercises 41

3.   Digital Convolution, Oscillators, and Windowing 43
3.1   Digital Convolution 43
3.1.1   Flip and Slip Method 43
3.2   DSP Sinusoidal Oscillator 48
3.3   Exercises 51

4.   Digital Filter Design Methods 53
4.1   Filter Types 53
4.2   Direct form 1 Filter 54
4.3   Direct form 2 Filter 55
4.4   The Transpose Filter 56
4.5   Cascade and Parallel Filter Realizations 57
4.5.1   Digital Filter Specification 57
4.6   The Bilinear Transform 60
4.6.1   Designing Digital Filters Using the Bilinear Transform Method 61
4.7   The Impulse-Invariant Filter Design Technique 64
4.7.1   Impulse Function Generation 65
4.7.2   Sampling the Impulse Response 66
4.7.3   Mapping from the s-Plane to the z-Plane 67
4.8   Truncating IIR Responses to Show Gibbs Effect 70
4.9   Designing Second-Order Filters Using the Impulse-Invariant Method 70
4.10   Windowing 72

4.10.1 Windows Plots 73

4.10.2 Windows Spectral Plots 75

4.11 Window Filter Design 75

4.11.1 Bartlett Window 76

4.11.2 The Sampled Impulse Response 76

4.12 Impulse Response of a Brick-Wall Filter 79

4.13 Designing Filters Using the Window Method 81

4.14 FIR Root-Raised Cosine Filter 83

4.14.1 Raised Cosine FIR Filter Design 85

4.14.2 Root-Raised Cosine FIR Filter Design 85

4.15 Exercises 86

5. **Digital Signal Processing Applications** 89

5.1 Telecommunication Applications 89

5.2 Quadrature Carrier Signals 89

5.3 Hilbert Transform 90

5.3.1 The Hilbert Impulse Response 92

5.3.2 The Hilbert Amplitude and Phase Responses 92

5.4 Single-Sideband Suppressed Carrier Modulation 93

5.5 Differentiator 94

5.6 Integrator 95

5.7 Multirate Systems: Decimation and Interpolation 95

5.8 Decimation 96

5.8.1 Example 100

5.8.2 Solution 101

5.9 Aliasing 101

5.10 Interpolation 102

5.11 Decimation and Interpolation for Noninteger Sampling Frequencies 103

5.12 Exercises 105

6. **Down-Sampling and Digital Receivers** 109

6.1 Receiver Design 109

6.2 RF Sampling 109

6.2.1 Down-Sampling a Passband Signal 110

6.2.2 Down-Sampling a Single-Sideband Signal 111

6.3 Digital Receiver 112

6.4 DSP and Music 114

6.4.1 Phasing Effect 118

6.5   Zero-Forcing Equalizer 120

    6.5.1   Three-Tap Zero-Forcing Equalizer 122

    6.5.2   Example 123

    6.5.3   Solution 123

6.6   Exercises 126

Appendix A: References 133

    Books 133

    Internet 133

Appendix B: Tables 135

Index 137

Author Biographies 141

Preface

In chapter 1 we examine the sampled signal and introduce the concept of digital frequency and examine DSP signals such as the unit step and unit impulse. The hierarchical method of drawing schematics is demonstrated using two types of delay unit used throughout the book. In chapter 2 we see how to express simple digital filters using difference equations (DE), and by means of the z-transform, convert the DE to a transfer function. We examine time-invariant systems for a range of filter types: low-pass, high-pass and bandpass. In chapter 3 we look at the important concept of convolution and demonstrate the usefulness of the 'log' command in Probe for correctly separating the displays to demonstrate the 'flip-n-slip' procedure. Digital oscillators, including quadrature carrier generation, are also examined in this chapter. In chapter 4, filter types such as the direct form types 1 and 2, and the transpose configuration are examined. We examine different filter design methods that include the bilinear transform, the impulse invariant, and window techniques. Included also is a treatment of the raised-cosine family of filters.

In chapter 5, a range of DSP applications are simulated and include the Hilbert transform, single sideband modulator using the Hilbert transform, quad oscillators, integrators and differentiators. Decimation and interpolation are simulated to demonstrate the usefulness of the multi-sampling environment. Chapter 6 looks at down-sampling and decimation as used in a treatment of digital receivers. Lastly, we look at some musical applications for DSP such as reverberation/echo using real-world signals imported into PSpice using the program Wav2Ascii. The zero-forcing equalizer is dealt with in a simplistic manner to show how the received signal is equalized.

ACKNOWLEDGEMENTS

I would like to thank Dr Eliathamby Ambikairajah (Ambi to his friends), associate Professor and Director of Academic Studies, University of New South Wales, Australia, who introduced me to the world of digital signal processing some years ago and did so in a very professional and caring manner. Two other people I would like to thank are: Professor Hussein Baher, a friend and colleague and a successful author of many textbooks on DSP. He gave me great encouragement and advice in getting published, and my son Lee Tobin who wrote a very useful program 'Wav2ASCII' create.

Please note for correct typographical representation, spaces have been included between numbers and dimensions i.e., 1 us.

CHAPTER 1

Introduction to Digital Signal Processing

1.1 SAMPLING

Digital signal processing (DSP) is used in most areas of electronics, telecommunications, and biomedical engineering. Devices such as compact disks, DVD write/readers, mobile phones, etc. are but a few of the many applications which contain digital signal processors. DSP principles are traditionally taught using Matlab$^{©}$ and Simulink$^{©}$, two excellent simulation software packages, but in this book we examine an alternative teaching method for understanding DSP fundamental principles. Fig. 1.1 shows the main signal processing in a typical digital signal processor. The first block is an antialiasing filter to ensure that frequencies greater than half the sampling frequency are attenuated such that the aliasing phenomenon is not a problem. The analog to digital converter (ADC) converts the sampled signal into binary form and the digital signal processor, similar to a microprocessor, implements algorithms such as filtering, musical effects such as echo, etc. A digital to analog converter (DAC) reconstitutes the original signal and a low-pass filter removes any components generated by the DAC.

FIGURE 1.1: DSP system.

1.2 ALIASING

Analog signals are continuous in time, whereas DSP signals are sampled in time $x(nT)$, with T being the period of the sampling signal and n an integer. The Claude Shannon–Harry Nyquist sampling theorem specifies the minimum sample rate (the Shannon–Nyquist rate) with respect to the maximum input signal frequency. Thus, a sampled analog signal, f_m, is recovered completely provided the sampling frequency $f_s \geq 2f_m$. To recover the original signal from a sampled signal, we must have a minimum of two samples per signal period. The input signal is band-limited before the sampling process by passing it through a low-pass, or bandpass filter.

If the analog signal is not band-limited to a maximum frequency of f_m, then low-frequency components above $f_s = 2f_m$ are aliased, or folded back, to introduce distortion in the recovered signal. The antialiasing low-pass filter attenuates frequencies greater than half the sampling frequency, reducing them to a level comparable to the quantization noise level so that we can ignore them.

1.2.1 The ADC Parameters

The ADC dynamic range is the ratio of maximum to minimum signal levels(in dB):

$$D = 20 \log V_{max} / V_{min}. \qquad (1.1)$$

Rounding off an analog signal amplitude to the nearest quantized value produces a noise voltage called *quantization noise*. We need to define a signal-to-quantization noise power ratio (SQNR) as the ratio of the signal present to the noise generated by the ADC. The 6 dB rule for an n-bit ADC, with a sinusoidal input signal occupying the maximum input, has a SQNR expressed as:

$$SQNR = 6n + 1.76. \qquad (1.2)$$

An 8-bit system will thus have an SQNR of 49.76 dB. Figure 1.2 shows how an ABM **MULT** part produces a sampled version of the input analog signal $x(t)$. The input wire segment name is labeled xt and not $x(t)$ or $x[t]$, as PSpice will not accept brackets or spaces in wire segment names. Set the transient parameters as follows: **Run to time** $= 10$ ms, and press **F11** to simulate.

Fig. 1.3 shows the sampled signal and the recovered signal from the **LOPASS** part (a low-pass filter).

Since T is constant we can write $x(nT)$ as $x(n)$ defined for integer values of n only. In the above simulation, the x-axis is time t but is normally shown as n in textbooks.

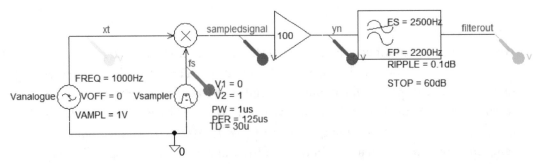

FIGURE 1.2: Sampling an analog signal.

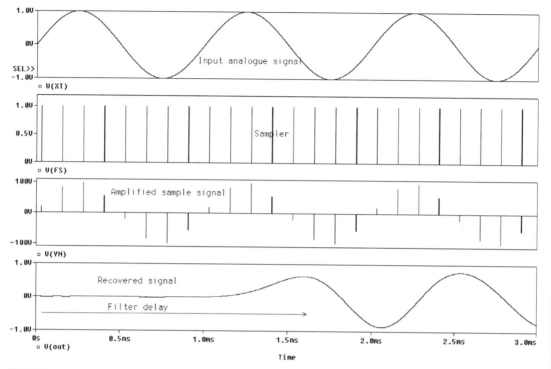

FIGURE 1.3: Sampled and recovered signals.

1.2.2   Digital Frequency

The instantaneous value of an analog sinusoidal signal is

$$x(t) = A \sin \omega_a t = A \sin 2\pi f_a t \ \ V. \tag{1.3}$$

A discrete equivalent of (1.3) is achieved by replacing continuous time, t, with sampled time, nT:

$$x(nT) = A \sin \omega_a nT = A \sin 2\pi f_a nT \ \ V. \tag{1.4}$$

We can write $x(nT)$ as $x(n)$ hence forth as T is constant for the systems under consideration. Comparing the angle of each sinusoid in (1.3) and (1.4) and substituting for the sampling frequency $f_s = 1/T$ in (1.4) gives an expression for the digital frequency:

$$x(n) = A \sin 2\pi n f_a / f_s \ \ V. \tag{1.5}$$

Here the digital frequency is defined as

$$\theta = 2\pi f_a / f_s. \tag{1.6}$$

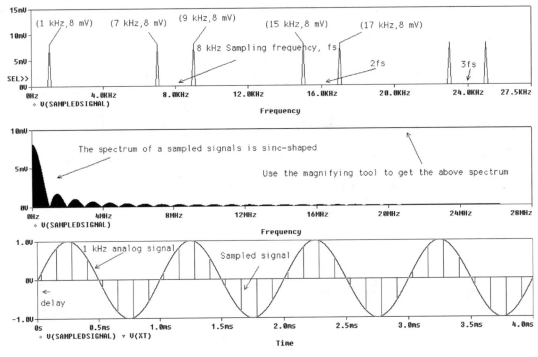

FIGURE 1.4: A sampled sine signal.

The digital frequency range is $-\pi \leq \theta \leq \pi$. Thus we may rewrite (1.5) as

$$x(n) = A \sin n\theta. \tag{1.7}$$

Fig. 1.4 shows an analog sinusoid superimposed on a sampled version of itself. The spectrum of the sampled signal is sinc-shaped but we may examine the spectrum in detail by copying the variable into a third plot, un-syncing the x-axis and pressing the FFT icon in PROBE. Note, however, that the sinc-shaped spectrum is all positive because of the FFT algorithm. Use the magnifying icon to examine the spectrum from 0 Hz. Note the sidebands are centered on the sampling frequency and also at multiples of the sampling frequency. Triangular-shaped spectral components may result if you choose too small a value for the **Run to time** transient parameter.

1.2.3   Digital Samples

A 1 kHz sinusoidal signal, when sampled at 8 kHz, produces a periodic digital frequency $\theta = 2\pi/8 = \pi/4$ and has a period N, where N is the smallest integer for which $x(n + N) = x(n)$. Thus,

$$y(n) = A \sin(n + N)\theta = A \sin n\theta \quad V. \tag{1.8}$$

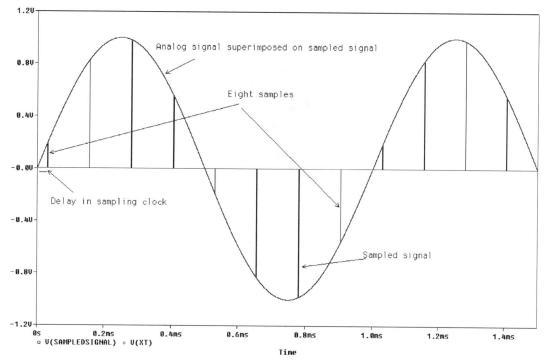

FIGURE 1.5: Signal sampled at $f_s = 8 f_a$.

This condition is satisfied for all n, if

$$N\theta = 2\pi k \Rightarrow N = \frac{2\pi k}{\theta} = \frac{f_s}{f_a} k. \qquad (1.9)$$

Here k is a constant. For example, the number of samples is calculated for an analog signal $f_a = 1$ kHz and sampling frequency $f_s = 8$ kHz, as

$$N = \frac{8\,\text{kHz}}{1\,\text{kHz}} = 8 \text{ samples.} \qquad (1.10)$$

A signal sampled at eight times the sampling frequency is shown superimposed on the original signal in Fig. 1.5. The sampling clock pulse was delayed by 30 us (this is the **TD** = 30 us parameter in the sampling **VPULSE** generator in Fig. 1.2) to show all eight samples; otherwise two sampling points would occur at the zero crossover point and would not be visible.

1.3 DELAY UNIT AND HIERARCHICAL BLOCKS

We will make use of hierarchical blocks quite a lot in DSP and the following description explains the technique for placing blocks. Select the hierarchical block from the right toolbar. From the **Place Hierarchical Block** menu, enter a name in the **Reference** box and in the

Place Hierarchical Block

Reference:

Timer

Primitive

- No
- Yes
- Default

Implementation

Implementation Type

Schematic View

Implementation name:

CR

Path and filename

FIGURE 1.6: Enter the block parameters as shown.

Implementation Type select **Schematic View** from the list. Fill in the **Implementation Name** of your choice and the **Path and Filename** as shown in Fig. 1.6 and press **OK**.

Draw the box using the left mouse button held down. Attach wires as necessary and from the toolbar menu select the pins icon as shown in Fig. 1.7 and fill in the parameters as shown.

Highlight the part with the left mouse button, **Rclick** the block and select **Descend Hierarchy** (short cut keystrokes: **shift+D** to get into block and **shift+A** to get back to main diagram) from the menu to display the inside of the block. In the new schematic area you should now see the input and output ports corresponding to the pin names on the block. If any pin tags are missing inside the block, just add them by selecting the **Place Port** tags icon. Draw the schematic and connect the input and output pins. The delay element is the main part in DSP and Fig. 1.8 shows two techniques for delaying a signal. The first method uses a correctly terminated transmission line to delay the signal, and the second delay uses a **Laplace** part. The transmission line delay is more robust but the **Laplace** part is useful for implementing high-order filters. A **PARAM** part specifies the transmission line delay = 125 us, and in the

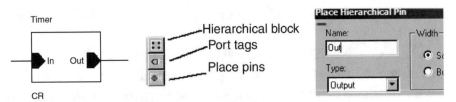

FIGURE 1.7: A hierarchical block with two pins attached.

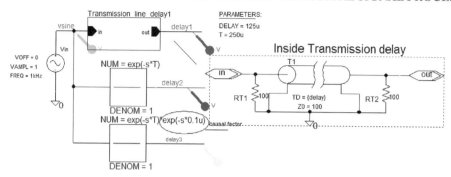

FIGURE 1.8: Two types of delays.

Laplace part the delay is $T = 250$ us (we could have used the same delay value for both but the different rate allows us to differentiate the signals when plotted in **PROBE**) [ref: 1 Appendix A].

1.3.1 Transmission Line Delay

Highlight the **T** part transmission line, **Rclick** and select **Edit Properties** to display the line parameters in the spreadsheet, part of which is shown in Fig. 1.9. Set the characteristic impedance Z_0 to 300 Ω and the delay **TD** to (you may specify any value as long as it is the same as the characteristic impedance, Z_0) 125 us. The input and output ports of the transmission line must be terminated with 300 Ω resistances to avoid reflections, which happens when the source and load resistances are not equal to the characteristic impedance. The Z_0 can be any value you choose. The delay is the inverse of the 8 kHz sampling frequency f_s, i.e., $T_s = 1/f_s = 1/8000$ Hz $= 125$ us (8000 Hz is the normal sampling rate used in the telecommunication industry where the sampling theorem states that for a sampling frequency of 8 kHz, the maximum input frequency must be less than 4 kHz).

Fig. 1.10 shows a transmission line where the delay, **TD**, is one of the transmission line part parameters to be entered. However, instead of typing in an actual value, we enter {**delay**} (brackets included). The actual delay of 125 us is now specified in the **PARAM** part by adding rows in the **PARAM** spreadsheet with **Name** = delay, and **Value** = 125 us (accessed by **Rclicking** and selecting **Edit Properties**).

Set the transient parameter **Run to time** to 5 ms and press **F11** to simulate.

TD	125u
Z0	300

FIGURE 1.9: Setting the transmission line parameters.

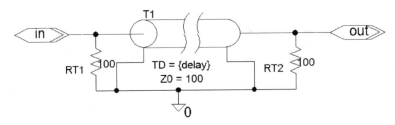

FIGURE 1.10: Inside the first delay block.

1.3.2 Laplace Delay

The z-transform, examined in Section 1.6, represents a delay as e^{-sT} where s is the complex frequency variable and T is the desired delay. The second delay type uses a **Laplace** part. This part normally expects a transfer function but here we enter exp($-s*T$) in the **NUM**erator box and 1 in the **DENOM**inator box (the Laplace part does not allow current, voltage, or time in the specification). T is specified in a **PARAM** part as **Name** = T and **Value** = 250 us entered in the spreadsheet as described in the previous section. The original and delayed signals are shown in Fig. 1.11.

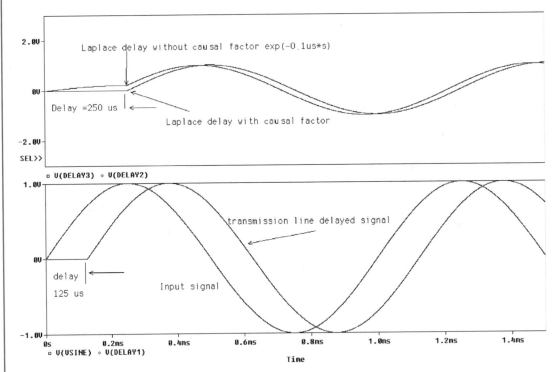

FIGURE 1.11: The original and delayed signals.

The **Laplace** part is useful but care must be taken to avoid noncausal type errors. It is also intensive in computational time and memory, so use transmission line delays where possible. A warning message might occur after simulation that tells us "WARNING PSpiceAD 02:21PM 6.65607 percent of E_LAPLACE impulse response is non-causal," and it should be delayed by at least "9.96094e-007 sec." This is solved by multiplying the numerator by $\exp(-s * 0.1u)$ to give an extra overall delay, i.e., the numerator is now $\exp(-s*T)* \exp(-s*0.1u)$. We use the **Append Waveform** in **PROBE** to display the delay with, and without, the causal factor. To append another **PROBE** display however, we need to modify the numerator and save the circuit with a different name, e.g., "Figure 1-011x.sch" and simulate again. From **File/Append Waveform(DAT)** select file "1-011x.prb."

1.4   DSP SIGNALS: UNIT STEP SIGNAL

We examine DSP signals in this section because it is important to know which signals to use and where to apply them when testing DSP circuits. Common test signals, such as the unit step and unit impulse, are examined and in particular the spectrum for the impulse signal is examined since it is such an important test signal. Fig. 1.12 shows how to produce a unit step function, $u(n)$, using a **VPULSE** part with period, **PER**, set to the sampling rate and the pulse width, **PW**, set to a smaller value.

The unit step signal in Fig. 1.13 is an infinite series of unit impulses delayed by the sampling time and is defined as

$$u(n) = \begin{Bmatrix} 1 & \text{for} & n \geq 0 \\ 0 & \text{for} & n \leq 0 \end{Bmatrix}. \tag{1.11}$$

1.4.1   Unit Impulse Signal

The unit impulse is the most important DSP signal for systems tests. A theoretical impulse or delta function is infinitely tall and thin, and contains all frequencies to infinity but is not

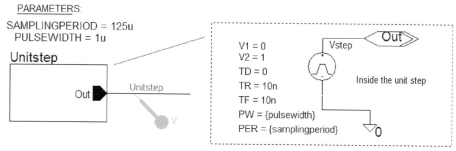

FIGURE 1.12: Unit step function generation.

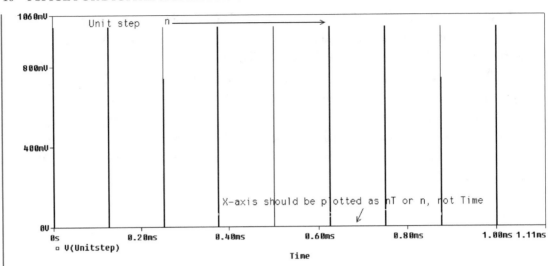

FIGURE 1.13: Unit step.

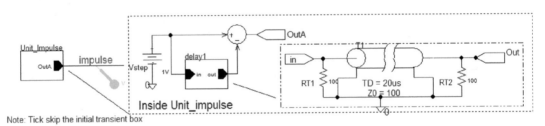

Note: Tick skip the initial transient box

FIGURE 1.14: Unit impulse generation.

realizable, so we use a unit impulse defined as

$$\delta(n) = \begin{cases} 1 & \text{at} \quad n = 0 \\ 0 & \text{elsewhere} \end{cases}. \tag{1.12}$$

It is important to understand that the impulse exists only at $n = 0$ and is zero elsewhere. The unit impulse may be expressed in terms of the step function as $\delta(n) = u(n) - u(n - 1)$. Thus, we generate a unit impulse by subtracting a delayed unit step from a unit step, leaving a signal whose width is the same as the width of the delay. Fig. 1.14 shows how a unit impulse, with amplitude of 1 V and pulse width of 20 us, is produced. Draw a hierarchical **Block** from the right toolbar menu and select the **Set Up Block**, the name of the circuit inside the block. Inside delay1 block is a correctly terminated line with a 100 Ω characteristic impedance. Subtraction is carried out using a **DIFF** part.

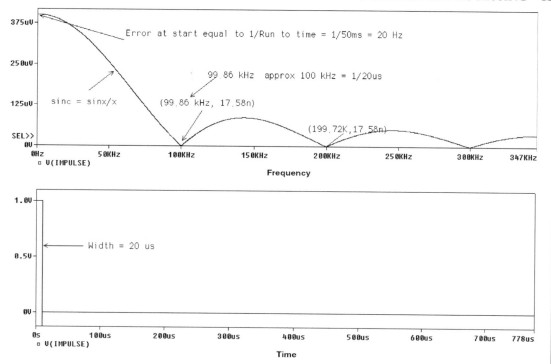

FIGURE 1.15: Unit impulse and sinc-shaped spectrum.

The delay is achieved by setting the delay parameters, TD, in a T part equal to 20 us. Subtracting the delayed step from the input step signal (a VDC part), yields a single impulse signal of width 20 us. The transient parameters are as follows: **Run to time** = 2 ms and **Maximum step size** = 0.1 us. Note that it is important to tick **Skip initial transient** in the **Transient Setup** menu. Press the **F11** key to simulate and display the unit impulse and spectrum shown in Fig. 1.15. The **Plot/Unsynchronize** function in **PROBE** allows a display of signals in the time and frequency domains simultaneously.

Press the **FFT** icon and, when the plot appears, apply the magnifying icon at the start of the sinc signal. This is necessary in order to display the first few null points where the signal goes through zero. The first null occurs at $\sin \pi f \tau = 0$, or $f = 10$ kHz $= 1/\tau = 1/100$ us and contains 90% of the total signal energy.

1.4.2   Decaying Exponential Signal

A decaying exponential signal is defined as

$$g(n) = \left\{ \begin{array}{ll} a^n & \text{at} \quad n \geq 0 \\ 0 & \text{elsewhere} \end{array} \right\}. \tag{1.13}$$

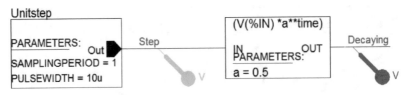

FIGURE 1.16: Exponentially decaying signal.

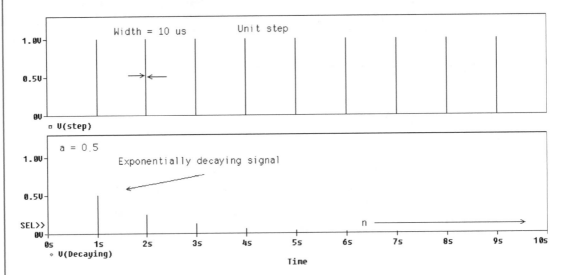

FIGURE 1.17: Decaying and unit step signals.

Here a is a fractional number entered in a **PARAM** part (0.5 in this example). The coefficient range is $0 < a < 1$. One **Param** part would suffice for all parameter definitions but we separate them here for simplicity. Fig. 1.16 shows how an exponentially decaying signal is achieved by applying a unit step to an **ABM1** part with the **Value** box EXP1 = (V(%IN) *a**time) where exponentiation is ** to implement the expression $a^n = a^{time}$.

The first block contains a unit step signal considered previously, and the second block defines the exponential signal a^{time}. Set the transient parameter **Run to time** to 10 s, press **F11** to simulate. Change the unit step parameters to observe the decaying nature of the signal shown in Fig. 1.17.

1.5   DIGITAL SEQUENCES

The digital signal in Fig. 1.18 consists of a series of weighted impulses delayed in time by replacing the variable n with $n - k$, where k is an integer greater than zero.

For example, $x(n - k)$ is the original signal $x(n)$ delayed by k samples. Note that an impulse advanced in time by k samples is $x(n + k)$. A sequence is written as the sum of scaled

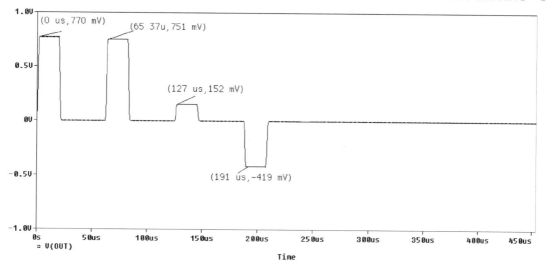

FIGURE 1.18: Typical DSP signal.

delayed unit impulses:

$$y(n) = a_0 x(0) + a_1 x(n-1) - a_2 x(n-2), \qquad (1.14)$$

where a_0, a_1, a_2, are the weighting coefficients of x and are called the *filter coefficients* (or taps). The output, $h(n)$, for a unit impulse, $\delta(n)$ input signal, is

$$h(n) = 0.77\delta(0) + 0.75\delta(n-1) + 0.152\delta(n-2) + -0.42\delta(n-2). \qquad (1.15)$$

An infinite sequence is expressed as

$$y(n) = \sum_{k=-\infty}^{\infty} a_k x(n-k). \qquad (1.16)$$

However, to mimic real-life signals we change the limits from $\pm$ infinity, to 0 to N finite samples.

$$y(n) = \sum_{k=0}^{N} a_k x(n-k). \qquad (1.17)$$

This equation is a form of the very important concept of convolution to be examined shortly.

1.6   THE z-TRANSFORM

The z-transform is examined in more detail in the next chapter but we introduce it briefly to simplify the analysis of delays (see Chapter 2, Section 2.2). The z-transform is defined as

$$X(z) = \sum_{n=-\infty}^{\infty} x(n)z^{-n}, \qquad (1.18)$$

where $z = e^{sT} = e^{j\omega T} = e^{j\theta}$, and T is the inverse of the sampling frequency. However, practical applications use the one-sided transform defined as

$$X(z) = \sum_{n=0}^{N} x(n)z^{-n}. \qquad (1.19)$$

1.6.1   Unit Impulse

The unit impulse exists only at $n = 0$ and can be used to test the frequency response of a system:

$$\delta(n) = \left\{ \begin{array}{ll} 1 & \text{at } \quad n = 0 \\ 0 & \text{elsewhere} \end{array} \right\}. \qquad (1.20)$$

Applying the z-transform to (1.20), so that the input signal is now

$$X(z) = \sum_{n=0}^{\infty} \delta(n)z^{-n} = \sum_{n=0}^{\infty} 1z^{-n} = z^{-0} + z^{-1} + z^{-2} = 1 + 0 + 0 \ldots . \qquad (1.21)$$

All other values are zero since the unit impulse has a value of 1 only at DC, i.e., $X(z) = \delta(0)$ is 1 at $n = 0$ only.

1.6.2   Unit Step

The unit step contains an infinite number of unit impulses and is defined as

$$u(n) = \left\{ \begin{array}{ll} 1 & \text{for } \quad n \geq 0 \\ 0 & \text{for } \quad n < 0 \end{array} \right\}. \qquad (1.22)$$

The z-transform of a unit step is

$$X(z) = \sum_{n=0}^{\infty} x(n)z^{-n} = \sum_{n=0}^{\infty} z^{-n} = z^{-0} + z^{-1} + z^{-2} \ldots = 1 + z^{-1} + z^{-2} \ldots . \qquad (1.23)$$

This infinite sum of delayed impulses has to be expressed in closed form. So consider the arithmetic series $S = a + ar + ar^2 + ar^3$. Multiplying by r yields $rS = ar + ar^2 + ar^3 + ar^4$

and subtracting these two equations yields

$$S - rS = a \Rightarrow S(-r) = a \Rightarrow S = \frac{a}{1-r}. \tag{1.24}$$

Thus (1.23) expressed in closed form using (1.24), with $a = 1$ and $r = z^{-1}$, is

$$\frac{1}{1-z^{-1}} x \frac{z}{z} = \frac{z}{z-1}. \tag{1.25}$$

1.6.3   Unit Delay

A unit impulse, $x(n)$, has an amplitude of 1 for $n = 0$ but zero elsewhere. When a unit impulse is delayed and transformed, the result is

$$X(z) = \sum_{k=0}^{1} x(n-k)z^{-n} = \sum_{k=0}^{1} 1z^{-k} = 1z^{-1} + 0 + 0... = z^{-1}.$$

1.7   EXERCISES

1. Investigate the spectrum of the various DSP signals examined.

2. A sinusoidal signal is sampled at a frequency f_s (note that $T_s = 1/f_s$). Deduce an expression for the digital frequency θ.

3. A DSP system is characterized by the difference function $y(n) = x(n) + 0 \cdot 5x(n-1)$. Apply the z-transform to this equation and obtain a transfer function.

4. Obtain the z-transform for the decaying exponential signal a^n.

CHAPTER 2

Difference Equations and the z-Transform

2.1 LINEAR TIME-INVARIANT SYSTEMS

A discrete time system is *linear* if an input signal, $x(n)$, is mapped onto the output signal, $y(n)$, as

$$a_1x_1(n) + a_2x_2(n) \to H \to b_1y_1(n) + b_2y_2(n). \qquad (2.1)$$

The constants a_1, a_2, b_1, b_2 are arbitrary and the system is linear because no extra components are generated by the system. The system is *time invariant* when both input and output signals are delayed by same amount, i.e.,

$$x(n - n_0) \to y(n - n_0). \qquad (2.2)$$

A system is linear time invariant (LTI) if both of these properties are met. A system is not time invariant when it uses different sampling frequencies throughout the system. A system has bounded-input bounded-output (BIBO) stability if a bounded (a limited amplitude signal) produces a bounded (limited) output sequence. A discrete time sequence is *causal* if, for all values of n less than zero, the output is zero. An LTI system, with an impulse response, $h(n)$, is causal, only if $h(n)$ is zero for $n < 0$. If the impulse response of an LTI system is of finite duration, the system is said to have a finite impulse response (FIR) but where infinite, it is an infinite impulse response (IIR).

2.2 DIFFERENCE EQUATIONS

A low-pass digital filter averages an input signal and a delayed signal and smoothes out, or integrates, any fast changing parts in the time signal. The input and output filter variables are related in a *difference equation* (DE) as

$$y(n) = x(n) + a_1x(n - 1). \qquad (2.3)$$

The output is the average (arithmetic mean) of the current and previous inputs. In the three-term averaging filter shown in Fig. 2.1, the input signal is multiplied by 1/3. The three-

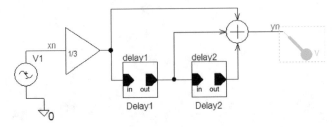

FIGURE 2.1: Three-term averaging filter.

input summer is a customized part but may be replaced by two, two-input sum parts if you cannot download my customized parts.

The difference equation for this system is

$$y(n) = 1/3[(x(n) + x(n-1) + x(n-2)].\tag{2.4}$$

2.3   DIGITAL FILTER CLASSIFICATION

Digital filters are classified by examining the impulse response. Apply a unit impulse signal (a single unit pulse existing at time $t = 0$ only) to the filter input and observe the output response. A filter is said to have a finite impulse response (FIR) filter when the number of terms in the response is finite, and an infinite impulse response (IIR) filter has an infinite number in the output (in practice, we cannot have an infinite number). These digital filters are characterized by a difference equation relating the input and output variables $x(n)$ and $y(n)$ as

$$y(n) = \sum_{k=0}^{M} a_k x(n-k) - \sum_{k=1}^{L} b_k y(n-k).\tag{2.5}$$

2.4   FIRST-ORDER FIR FILTER

A filter may contain feed-forward or feedback terms, or both. A *recursive* filter is one whose output depends on *past* output values, whereas a nonrecursive FIR filter has an output that depends on present input values only. Consider the first-order nonrecursive FIR filter in Fig. 2.2. Summation is implemented using a **SUM** part to sum the feed-forward direct $x(n)$ signal and the delayed signals $ax(n-1)$, where the weighting factor a is a fractional coefficient called the *filter coefficient* (also called a filter tap). The coefficient is not given a value but labeled. The actual value is specified as 0.5 in a **PARAM** part. Since the evaluation PSpice has no delay part, we must use a correctly terminated transmission line (part name **T**). The input wire segments are labeled using the **Place Net Alias** icon from the right toolbar. Use simple names and do not include spaces or brackets such as $x[n]$, or $x(n)$. The **AC Analysis** parameters are as follows: **Start Frequency** $= 1$, **End Frequency** $= 10$ kHz, **Points/Decade** $= 1000$ and **F11** to simulate.

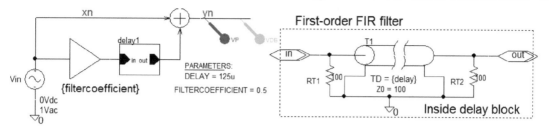

FIGURE 2.2: FIR filter with $a = 0.5$.

The low-frequency amplitude response in Fig. 2.3 is plotted on a log scale. Note however, there is also a mirror image frequency response that is present in all sampled signal spectra. The response repeats to infinity with each response centered on multiples of the sampling frequency, i.e. fs, 2fs, etc.

However, we are interested in the region from 10 Hz to 4000 Hz only (4000 Hz is half the sampling frequency sometimes shown as π). Fig. 2.4 shows the frequency response on a linear scale. The phase response is also shown to the left of the amplitude response and is linear over the passband region.

The output $y(n)$ is the sum of the signals into the summer and is called a difference equation:

$$y(n) = x(n) + 0 \cdot 5x(n-1). \tag{2.6}$$

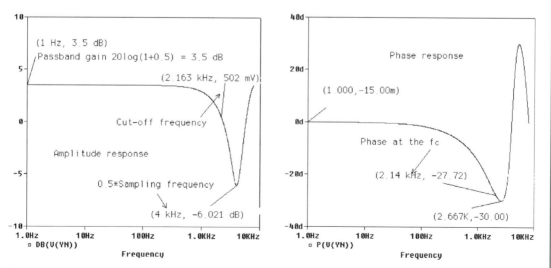

FIGURE 2.3: Frequency response showing the mirror image.

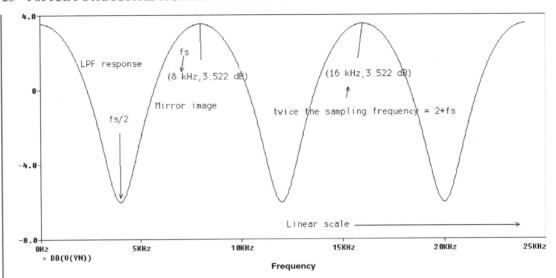

FIGURE 2.4: Frequency response with linear x-axis.

Changing the sign of the coefficient to $a = -0.5$ changes the low-pass integrating filter into a high-pass differentiating filter.

2.4.1 z-Transforms and Difference Equations

The z-transform enables us to obtain a transfer function from the difference equation. The z-transform was defined in Chapter 1 but we define it again as

$$X(z) = \sum_{n=-\infty}^{\infty} x(n)z^{-n}, \qquad (2.7)$$

where $z = e^{sT}$. The one-sided transform is defined as

$$X(z) = \sum_{n=0}^{N} x(n)z^{-n}. \qquad (2.8)$$

A difference equation is expressed in general terms as

$$Y(z) = \sum_{k=0}^{M} a_k X(z) z^{-k} - \sum_{k=1}^{L} b_k Y(z) z^{-k}. \qquad (2.9)$$

This covers the case of IIR and FIR filters but FIR filters are produced by making the second part zero. The transfer function is the ratio of $Y(z)$ to $X(z)$ as

$$H(z) = \frac{Y(z)}{X(z)} = \frac{\sum_{k=0}^{M} a_k z^{-k}}{1 + \sum_{k=1}^{L} b_k z^{-k}}. \qquad (2.10)$$

Thus, DSP filters contain combinations of delays, filter coefficients and summers. Adding the sampled input signal, $x(n)$, to a delayed scaled version, $a(k)x(n-1)$, produces low-pass filtering. The filter order is determined by the number of delay units in the filter, so two delays is a second-order filter. To obtain a transfer function we must z-transform all variables in n so that the difference equation is expressed in terms of z. For example, a single time delay is represented in the z-domain by multiplying the signal by z^{-1}, so (2.6) becomes

$$Y(z) = X(z) + 0.5 X(z)z^{-1} = X(z)(1 + 0.5z^{-1}). \qquad (2.11)$$

The transfer function is derived by collecting similar terms (2.11) and writing

$$H(z) = \frac{Y(z)}{X(z)} = 1 + 0.5z^{-1}. \qquad (2.12)$$

The passband gain at DC is calculated by substituting for $s = 0$, i.e., $z^{-1} = e^{-sT} = e^{-j2\pi fT} = e^{-j2\pi 0T} = e^{-0} = 1$. Thus, we may use (2.11) to express the transfer function passband gain in dB as $20\log(1 + 0.5) = 3.5$ dB.

2.5   POLE-ZERO CONSTELLATION AND BIBO STABILITY

The stable left-hand side of the s-plane is mapped into the unit circle of the z-plane. It is a unit circle since the magnitude of the z-transform is 1, i.e., $z = 1.\exp(sT)$. To investigate DSP systems stability, we must consider the poles and zeros of the transfer function. A pole is the value of z in the transfer function that makes it infinite, and a zero is the value of z that makes the transfer function zero. Consider an FIR first-order filter difference function:

$$y(n) = x(n) + 0 \cdot 5x(n-1).$$

z-transforming this equation and expressing the transfer function, $H(z)$, yields

$$Y(z) = X(z) + 0.5X(z)z^{-1} \Rightarrow H(z) = \frac{Y(z)}{X(z)} = 1 + 0.5z^{-1} = \frac{z + 0.5}{z}.$$

Here $(1 + 0.5z^{-1})$ is manipulated into a suitable form by multiplying $H(z)$ above and below by z. The transfer function is zero when $z = -0.5$. Hence, we say that there is a zero at

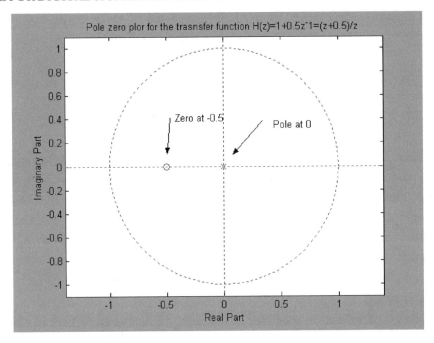

FIGURE 2.5: A pole–zero plot.

this value. Similarly, there is a pole at $z = 0$, where the transfer function tends toward infinity since we are attempting to divide by zero (use the L'Hopitals rule). Open Matlab and define the denominator coefficients as $\mathbf{a} = \mathbf{1}$, and the numerator coefficients as $\mathbf{b} = [\mathbf{1}, \mathbf{0.5}]$. To plot the pole–zeros as in Fig. 2.5, use the mfile **zplane (b, a)**.

Poles in the z-plane are represented by the symbol x and zeros as small circles. Double zeros are denoted by a small box around the zero/pole (or a little 2 beside the location). A system is stable if the poles of a transfer function are located within the unit circle. If the poles are located on the unit circle, the system has marginal stability, such as in an oscillator. However, the system is unstable if poles are located outside the unit circle and does not meet the BIBO criterion. A system with poles within the unit circle is said to be BIBO stable. The zeros modify the response but are not directly responsible for system stability.

2.6   CUT-OFF FREQUENCY

The cut-off frequency is that frequency where the gain is reduced to 0.707 times the passband gain. This is equivalent to the passband gain/$\sqrt{2}$. To derive an expression for the -3 dB cut-off frequency, we must first convert the transfer function, as a function of z, to a function of θ, the digital frequency. Applying Euler's formula $e^{\pm jx} = \cos x \pm j \sin x$ to each z term in the transfer function $H(z) = Y(z)/X(z) = 1 + a_0 z^{-1}$ yields

$$H(z)\big|_{e^s T} = H(e^{j\omega T}) = 1 + a_0 e^{-j\theta} = (1 + a_0 \cos\theta) - j(a_0 \sin\theta). \qquad (2.13)$$

Squaring the real and imaginary parts of (2.13) yields the transfer function in magnitude-squared form as

$$|H(\theta)|^2 = (1 + a_0 \cos\theta)^2 + (a_0 \sin\theta)^2. \qquad (2.14)$$

Equating (2.14) to the square of the (passband gain)$/\sqrt{2}$ yields an expression for the cut-off frequency. The passband gain is $1 + a_0$, since at zero frequency (DC), $z^{-1} = e^{-sT} = e^{-j\omega T} = e^{-0T} = 1$:

$$1 + 2a_0 \cos\theta + a_0^2(\cos^2\theta + \sin^2\theta) = (\text{passband gain}/\sqrt{2})^2 = (1 + a_0)^2/2. \qquad (2.15)$$

Note: $(\cos^2\theta + \sin^2\theta) = 1$.

Rearranging (2.15) and substituting for the digital frequency $\theta = 2\pi f_c/f_s$ yields the cut-off frequency f_c:

$$f_c = \frac{f_s}{2\pi} \cos^{-1}\left\{ \frac{(2a_0 - 1 - a_0^2)}{4a_0} \right\}. \qquad (2.16)$$

The cut-off frequency is thus dependent on the filter coefficient and the sampling rate. To investigate the effect of the coefficient value on the cut-off frequency, include a **PARAM** part in the first-order LPF as shown in Fig. 2.6 and enter the required value.

Select the **PARAM** part, **Rclick** and select **Edit Properties**. Click **New Row** and in **Add New Row** fill in **Name = filtercoef**, and **Value** = 0.1. The filter coefficient is represented by a **GAIN** part. **Lclick** this part and enter the gain value as {filtercoef}. The curly braces must be included! From the **Analysis** menu, select **Parametric,** and set **Swept Var. Type** to **Global Parameter**. In the **Name** box enter **filtercoef**, and **Start Value** = 0.1, **End Value** = 0.9, and **Points/Octave** = 0.1. Set **AC Sweep** and **Noise Analysis** parameters: AC sweep type to **Logarithmic/Decade, Start Frequency** = 1 Hz, **End Frequency** = 4 kHz, **Points/Decade** =

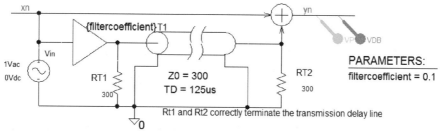

FIGURE 2.6: A first-order low-pass FIR filter.

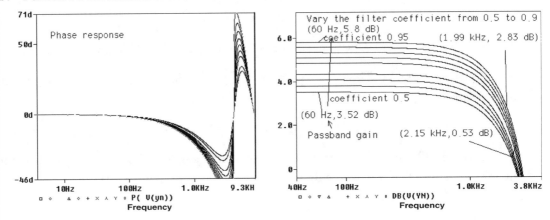

FIGURE 2.7: Varying the filter coefficient produces a range of f_c.

1001. Press **F11** to produce the response shown in Fig. 2.7. The filter coefficient is varied from 0.1 to 0.9 in steps of 0.1, and so the passband gain also varies since this is equal to $20 \log(1+\text{filtercoef})$ dB. For a filter coefficient of 0.9, the passband gain is $20 \log(1 + 0.9) = 5.5$ dB, and for a coefficient of 0.1, the passband gain is $20 \log(1 - 1) = 0.82$ dB.

2.7 INFINITE IMPULSE RESPONSE FILTER

A filter with feedback from the output, rather than feed-forward from the input, as in the FIR filter, is called a recursive infinite impulse response filter (IIR). A *recursive* filter is one where the present output depends on past output values. In this example, there are two terms: the original input signal $x(n)$, and a delayed output signal $y(n - 1)$ multiplied by the filter coefficient b. The output $y(n)$ is the sum of the inputs to the summer, i.e.,

$$y(n) = x(n) + b_0 y(n - 1). \tag{2.17}$$

2.7.1 First-Order Low-Pass IIR Filter

Draw the first-order IIR LPF schematic shown in Fig. 2.8. Delay1 block contains a transmission line **T** (or a **Laplace** part with a transfer function $\exp(-s*T)$) :

$$Y(z) = X(z) + b_0 Y(z)z^{-1} \Rightarrow Y(z)(1 - b_0 z^{-1}) = X(z). \tag{2.18}$$

Equation (2.18) is manipulated to yield a transfer function with a coefficient $b_0 = 0.9$:

$$H(z) = \frac{Y(z)}{X(z)} = \frac{1}{1 - 0.9z^{-1}}. \tag{2.19}$$

To determine the LPF passband gain (the gain at DC, or zero frequency), we substitute zero for the value of s, i.e., $e^{-jsT} = z^{-0} = 1$. Expressing (2.19) in decibels it is $20^* \log(1/$

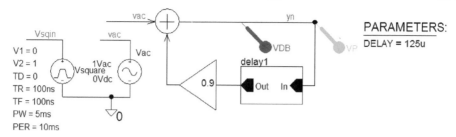

FIGURE 2.8: A first-order IIR low-pass filter.

$(1 - 0.9) = 20$ dB. Substituting for $z^{-1} = e^{-jsT} = e^{-j\theta}$ in (2.19), and using Euler's identity, yields the transfer function in cosine and sin terms (real and imaginary components) as

$$H(e^{-j\theta}) = H(\theta) = \frac{1}{1 - b_0 e^{-j\theta}} = \frac{1}{1 - b_0 \cos\theta + jb_0 \sin\theta}. \qquad (2.20)$$

The passband gain (the gain at DC for an LPF) is obtained by making the angle zero:

$$H(\theta) = \frac{1}{1 - b_0 e^{-j\theta}} = \frac{1}{1 - b_0 e^{-j0}} = \frac{1}{1 - b_0}. \qquad (2.21)$$

Squaring (2.21) yields the transfer function in magnitude-squared form:

$$\left|H(\theta)\right|^2 = \frac{1}{(1 - b_0 \cos\theta)^2 + (b_0 \sin\theta)^2}. \qquad (2.22)$$

Equate (2.22) to the (passband gain)/$\sqrt{2}$ squared:

$$\frac{1}{(1 - b_0 \cos\theta)^2 + (b_0 \sin\theta)^2} = \left\{\frac{\text{passband gain}}{\sqrt{2}}\right\}^2 = \left\{\frac{1/(1 - b_0)}{\sqrt{2}}\right\}^2. \qquad (2.23)$$

2.7.2   Cut-off Frequency for the IIR Filter

Rearranging (2.23) and substituting $2\pi f_a/f_s = \theta$ yields an expression for the cut-off frequency:

$$f_c = \frac{f_s}{2\pi} \cos^{-1}\left\{\frac{4b_0 - 1 - b_0^2}{2b_0}\right\}. \qquad (2.24)$$

Substituting for $b_0 = 0.9$ and $f_s = 8000$ Hz yields the -3 dB cut-off frequency as 134 Hz. Set **AC Sweep** and **Noise Analysis** parameters: AC sweep type to **Logarithmic/Decade,Start Frequency** = 1 Hz, **End Frequency** = 4 kHz, **Points/Decade** = 1001. Simulate with **F11** to produce the response in Fig. 2.9. Since the passband gain is 20 dB, we measure the cut-off frequency at the location where the output is 3 dB below this (17 dB) and is found to be 134 Hz.

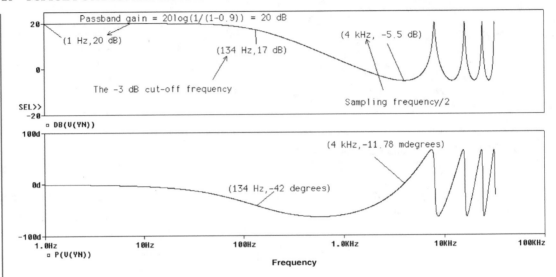

FIGURE 2.9: IIR LPF response.

2.7.3   IIR Phase Response

We investigate phase and group delay responses for first-order FIR and IIR filters by placing a phase marker on the output and changing the x-axis to linear. There is a marker for group delay as well.

2.8   HIGH-PASS IIR FILTER

The IIR recursive high-pass filter, with a -0.5 filter coefficient, is shown in Fig. 2.10, where the delay uses a transmission line or the **Laplace** part whose transfer function is e^{-sT}. If you use a Laplace delay, then you must reverse the **Laplace** part by selecting the part, **Rclicking** and selecting **Mirror Horizontally**—a necessary step since the feedback signal path direction is from output to input. The sampling period ($T = 125$ us) is defined in the **PARAM** part.

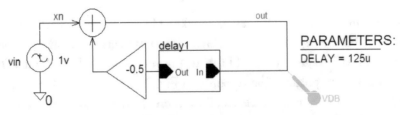

FIGURE 2.10: A first-order IIR high-pass filter.

2.8.1   The −3 dB Cut-Off Frequency

The difference equation for this filter is $y(n) = x(n) - 0 \cdot 5y(n-1)$, so the IIR high-pass filter transfer function is

$$H(z) = \frac{Y(z)}{X(z)} = \frac{1}{1 + 0.5z^{-1}}. \qquad (2.25)$$

We must derive an expression for the cut-off frequency, f_c, in a similar manner to that carried out for the LPF. We must equate the transfer function to the passband gain divided by the square root of 2 ($1/\sqrt{2} = 0.707$ is the definition of the cut-off frequency). Substitute $e^{-j\theta}$ in (2.25), and apply Euler's identity to express the *TF* in cosine and sin terms (the real and imaginary components) as

$$H(e^{j\omega T}) = H(\theta) = \frac{1}{1 + b_0 e^{-j\theta}} = \frac{1}{1 + b_0 \cos\theta - jb_0 \sin\theta}. \qquad (2.26)$$

Squaring (2.26) yields the *TF* in magnitude-squared form:

$$\left| H(\theta) \right|^2 = \frac{1}{(1 + b_0 \cos\theta)^2 + (b_0 \sin\theta)^2}. \qquad (2.27)$$

Equate (2.27) to the (passband gain)/$\sqrt{2}$ squared:

$$\frac{1}{(1 + b_0 \cos\theta)^2 + (b_0 \sin\theta)^2} = \left[\frac{\text{passband gain}}{\sqrt{2}} \right]^2. \qquad (2.28)$$

2.8.2   Passband Gain

The passband gain occurs at a frequency $f_a = f_{s/2}$, which is the digital frequency $2\pi f_{s/2}/f_s = \pi$ (an angle of 180° so that $\cos(\pi) = -1$ and $\sin(\pi) = 0$). The passband gain is therefore

$$\left. \left| H(\theta) \right| \right|_{\theta=\pi} = \frac{1}{1 + b_0 \cos\pi - jb_0 \sin\pi} = \frac{1}{1 - b_0 + 0} = \frac{1}{1 - 0.5} = 2. \qquad (2.29)$$

Note: When substituting in the coefficient value into (2.29), use the magnitude only since the sign was taken into account when obtaining the transfer function.

Thus, the passband gain in dB is $20 \log(2) = 6$ dB. Substituting 2 into (2.28) yields $1 = 2(1 + b_0 \cos\theta)^2 + (b_0 \sin\theta)^2$, so that

$$2(1 + 2b_0 \cos\theta + b_0^2 \cos^2\theta + b_0^2 \sin^2\theta) = 2 + 4b_0 \cos\theta + 2b_0^2). \qquad (2.30)$$

Manipulating (2.30) yields

$$1 = 2 + 4b_0 \cos\theta + 2b_0^2. \qquad (2.31)$$

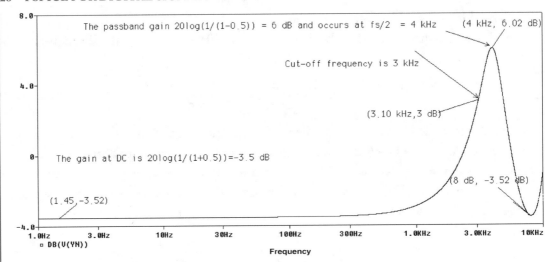

FIGURE 2.11: IIR high-pass filter frequency response with $b = -0.5$.

An expression for the cut-off frequency is obtained by substituting for the digital frequency $\theta = 2\pi f_a/f_s$ and rearranging (2.31):

$$f_c = \frac{f_s}{2\pi} \cos^{-1}\left\{-\left(\frac{1 + 2b_0^2}{4b_0}\right)\right\}. \qquad (2.32)$$

Substituting for $b_0 = 0.5$ and $f_s = 8000$ Hz yields a 3 kHz cut-off frequency. Set **AC Analysis** range as 1 Hz to 10 kHz, and press **F11** to produce the IIR HPF frequency response shown in Fig. 2.11. The gain at DC (frequency = 0) is calculated by making the delay term 1, i.e., $z^{-0} = 1$, so the transfer function is $1/(1 + b_0 \cos(0) + b_0 \sin(0)) = 1/(1 + 0.5)$. Expressed in decibels it is $20^* \log(1/(1 + 0.5) = -3.5$ dB.

This looks like a bandpass response but do not forget that the response is from DC to $f_s/2 = 4000$ Hz, before it wraps itself around giving a mirror image, as in all sampled signal systems responses. Vary the filter coefficient using the **Parametric** option from the **Analysis Setup** menu and define the initial coefficient value in a **PARAM** part. Observe the effect on the response.

2.9   STEP RESPONSE

The output $y(n)$ for a step input signal $u(n)$ is called the step response. Apply the step signal tool created to the FIR filter shown in Fig. 2.12.

We could, of course, just apply a **VPULSE** with the parameters as shown to produce the input and the weighted delayed input signals as plotted in Fig. 2.13.

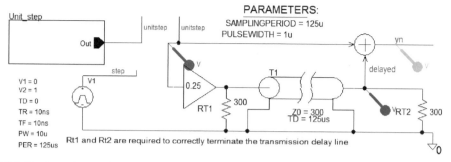

FIGURE 2.12: Impulse testing.

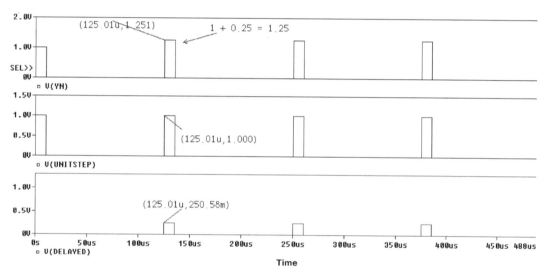

FIGURE 2.13: Step response of the FIR filter.

2.10   IMPULSE TESTING

The output $y(n)$ for an impulse input signal $x(n) = \delta(0)$ is called the impulse response, where $y(n)$ is replaced by the symbol $h(n)$. The impulse response characterizes a system since an ideal impulse signal contains all frequencies from DC to infinity. Thus, applying an impulse is the same as connecting a frequency generator and sweeping the frequency over a large frequency range. However, we need to use the **FFT** icon to get the frequency response. A *recursive system* is one whose output is dependent on past output values as shown in Fig. 2.14. If the input analog signal is sampled at 8.0 kHz, obtain the impulse response by calculating values for the impulse response $h(n)$, for $n = 0, 1, 2, 3, 4$. A unit impulse function is approximated as a pulse train whose period (**PER**) is very long, and a very short 20 us pulse width (**PW**). These values are a compromise between short simulation time and accuracy.

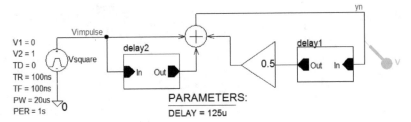

FIGURE 2.14: First-order filter.

The IIR filter difference equation is

$$y(n) = x(n) + x(n-1) + 0.5y(n-1). \qquad (2.33)$$

Applying a unit impulse $\delta(0)$ to the input produces an output $h(n)$—the impulse response:

$$h(n) = \delta(n) + \delta(n-1) + 0.5h(n-1). \qquad (2.34)$$

Compute the $h(n)$ coefficients by substituting values for $n = 0, 1, 2, 3$, and 4 in (2.34):

$$
\begin{aligned}
n = 0 \qquad & h(0) = \delta(0) + \delta(0-1) + 0.5h(0-1) = 1 + 0 + 0.5(0) = \mathbf{1} \\
n = 1 \qquad & h(1) = \delta(1) + \delta(0) + 0.5h(0) = 0 + 1 + 0.5(1) = \mathbf{1.5} \\
n = 2 \qquad & h(2) = \delta(2) + \delta(1) + 0.5h(1) = 0 + 0 + 0.5(1-5) = \mathbf{0.75} \\
n = 3 \qquad & h(3) = 0 + 0 + 0.5(0.75) = \mathbf{0.375} \\
n = 4 \qquad & h(4) = 0\ 0 + 0.5(0.375) = \mathbf{0.1875}.
\end{aligned}
$$

Compare these results to those shown in Fig. 2.15. Note that the **Maximum step size** in the **T̲ransient Analysis** setup should be 10 ns or less and *s* larger value will result in incorrect plotted impulse values.

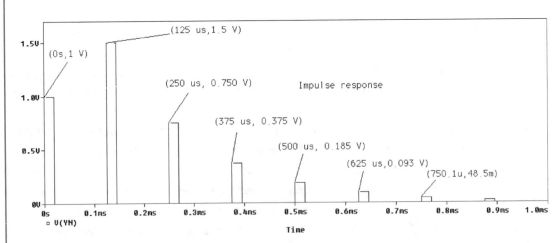

FIGURE 2.15: The impulse response.

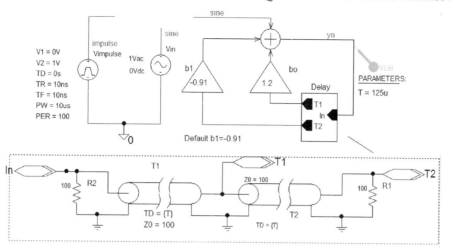

FIGURE 2.16: A recursive bandpass filter.

2.11   BANDPASS IIR DIGITAL FILTER

A difference equation for the second-order bandpass filter in Fig. 2.16 is

$$y(n) = x(n) + b_0 y(n-1) + b_1 y(n-2) = x(n) + 1.2 y(n-1) - 0.91 y(n-2). \quad (2.35)$$

Be careful when terminating cascading transmission lines; **Resistances are only required at the input and output for matching purpose, and not at the junction of the two lines where the output and input impedances are the same.** Applying the z-transform to (2.35) yields the second-order bandpass filter transfer function:

$$\frac{Y(z)}{X(z)} = \frac{1}{1 - b_0 z^{-1} - b_1 z^{-2}} = \frac{1}{1 - 1 \cdot 2 z^{-1} + 0.91 z^{-2}} = \frac{z^2}{z^2 - 1 \cdot 2z + 0.91}. \quad (2.36)$$

2.11.1 Bandpass Pole–Zero Plot

In order to draw the pole–zero plot, we must find the roots of the denominator using the expression below, where $a = 1$, $b = -1.2 = b_0$, and $c = 0.91 = b_1$:

$$\frac{-b \pm \sqrt{b^2 - 4ac}}{2a} = \frac{-(-1 \cdot 2) \pm \sqrt{1 \cdot 44 - 3 \cdot 64}}{2} = 0 \cdot 6 \pm j0 \cdot 74. \quad (2.37)$$

When the real roots of a second-order polynomial are positive, the polynomial is written with the root entered as a negative number, so that in this example it is -0.6, and not $+0.6$. In Matlab, if we define the coefficient matrix as $c = [1, -1.2, 0.9]$, then the mfile **roots(c)** gives us the roots as $0.6 \pm j0.7416$. When substituting these roots back into (2.36), we must put a negative sign in front:

$$H(z) = \frac{z^2}{(z - 0 \cdot 6 - j0 \cdot 74)(z - 0 \cdot 6 + j0 \cdot 74)}. \qquad (2.38)$$

A general form for (2.38) is

$$H(z) = \frac{1}{1 + b_0 z^{-1} + b_1 z^{-2}}, \qquad (2.39)$$

where $b_0 = 1.2$ and $b_1 = -0.91$. The roots of (2.39) are calculated as

$$\frac{-b_0 \pm \sqrt{b_0^2 - 4b_1}}{2} = \frac{-b_0 \pm \sqrt{b_0^2 - 4b_1}}{2} = \frac{-b_0}{2} \pm j\frac{\sqrt{4b_1 - b_0^2}}{2}. \qquad (2.40)$$

Rearranged the square root by introducing the imaginary number $j = \sqrt{(-1)}$. Plot the constellation with **zplane(1, c)**. The horizontal component from the pole–zero display in Fig. 2.17 gives us a value for the first coefficient.

2.11.2 The −3 dB Cut-Off Frequency

We need to get an expression for the cut-off frequency in terms of the filter coefficients and sampling frequency:

$$\frac{b_0}{2} = r \cos \theta_0 \Rightarrow b_0 = 2r \cos \theta_0. \qquad (2.41)$$

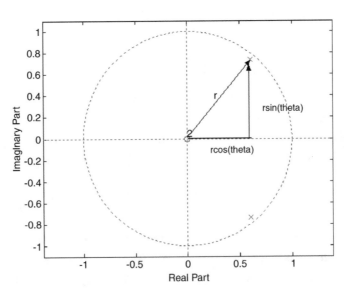

FIGURE 2.17: Pole–zero plot.

Also

$$\frac{\sqrt{4b_1 - b_0^2}}{2} = r\sin\theta_0 \Rightarrow 4b_1 - b_0^2 = 4r^2\sin^2\theta_0. \tag{2.42}$$

Substituting b_0 from (2.41) into (2.42) yields

$$4b_1 - 4r^2\cos^2\theta_0 = 4r^2\sin^2\theta_0 \Rightarrow 4b_1 = 4r^2(\sin^2\theta_0 + \cos^2\theta_0) = 4r^2. \tag{2.43}$$

Since $\sin^2\theta_0 + \cos^2\theta_0 = 1$, hence (2.43) gives us a value for the second filter coefficient:

$$4b_1 = 4r^2 \Rightarrow b_1 = r^2. \tag{2.44}$$

Substitute the two coefficients (2.41) and (2.44) into (2.39) to yield

$$H(z) = \frac{1}{1 + b_0 z^{-1} + b_1 z^{-2}} = \frac{1}{1 - (2r\cos\theta_0)z^{-1} + r^2 z^{-2}}. \tag{2.45}$$

The resonant digital frequency is $\theta_0 = 2\pi f_0/f_s$. Substituting the second coefficient, defined in (2.44), into (2.47), gives the resonant digital frequency in terms of the filter coefficients:

$$\cos\theta_0 = \frac{b_0}{2\sqrt{b_1}} \Rightarrow \theta_0 = 2\pi\frac{f_0}{f_s} = \cos^{-1}\left(\frac{b_0}{2\sqrt{b_1}}\right). \tag{2.46}$$

The final expression for the center frequency with values substituted yields

$$f_0 = \frac{f_s}{2\pi}\cos^{-1}\left(\frac{b_0}{2\sqrt{b_1}}\right) = \frac{8000}{2\pi}\cos^{-1}\left(\frac{1.2}{2\sqrt{0.91}}\right) = 1133 \text{ Hz}. \tag{2.47}$$

Change the input wire segment name to sine and set **AC Sweep** and **Noise Analysis** parameters to **Points/Decade** $= 1001$, **Start Frequency** $= 100$, **End Frequency** $= 4$ kHz. Press **F11** to simulate and produce the bandpass frequency response in Fig. 2.18. Use two cursors to measure the -3 dB bandwidth.

An alternative derivation for the cut-off frequency is

$$\frac{r\sin\theta_0}{r\cos\theta_0} = \tan\theta_0 = \frac{(\sqrt{4b_1 - b_0^2})/2}{b_0/2} = \frac{\sqrt{4b_1 - b_0^2}}{b_0} \Rightarrow \theta_0 = \tan^{-1}\left(\frac{\sqrt{4b_1 - b_0^2}}{b_0}\right) \tag{2.48}$$

$$\Rightarrow f_0 = \frac{f_s}{2\pi}\tan^{-1}\left(\frac{\sqrt{4b_1 - b_0^2}}{b_0}\right) = \frac{8000}{2\pi}\tan^{-1}\left(\frac{\sqrt{4.91 - (1.2)^2}}{1.2}\right) = 1133 \text{ Hz}. \tag{2.49}$$

The bandwidth is measured accurately using a goal function selected from **Trace Evaluate measurement**. Select **Bandwidth(V(BANDPASSOUT), 3)** from the list and click **OK**. Type

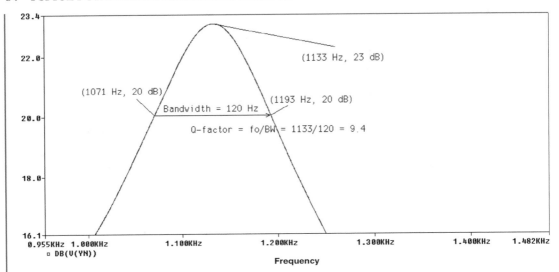

FIGURE 2.18: Bandpass frequency response.

the output variable **V(Bandpassout)** in the white box, and in the other box, enter 3:

$$H(z) = \frac{1}{1 - 1 \cdot 2z^{-1} + 0.91z^{-2}}$$

$$= \frac{1}{1 - 1 \cdot 2 \cos\theta + j1.2 \sin\theta + 0.91 \cos 2\theta - j0.91 \sin 2\theta} \quad (2.50)$$

$$H(z) = \frac{1}{1 - 1 \cdot 2 \cos\theta + 0.91 \cos 2\theta + j(1.2 \sin\theta - 0.91 \sin 2\theta)}. \quad (2.51)$$

The imaginary term (the j term) at resonance is not necessarily zero, so we need to compute the gain at resonance from the absolute value of (2.51) as $20 \log |H(z)|$:

$$|H(z)|_{dB}^{f_0} = 20 \log \left(\frac{1}{[(1 - 1 \cdot 2 \cos\theta + 0.91 \cos 2\theta)^2 + (1.2 \sin\theta - 0.91 \sin 2\theta)^2]^{0.5}} \right) = 23 \text{ dB}.$$
$$(2.52)$$

The quality factor is calculated from the resonant frequency and bandwidth as

$$Q = \frac{f_0}{BW} = \frac{1133}{120} = 9.4. \quad (2.53)$$

Measure the bandwidth and resonant frequency from the frequency response, and hence calculate the Q-factor using (2.53). To change the resonant frequency to 3 kHz, we need to recalculate the coefficients. To reduce the degrees of freedom, use the b_0 value but calculate a new value for b_1. Measure the maximum gain, BW, and Q-factor.

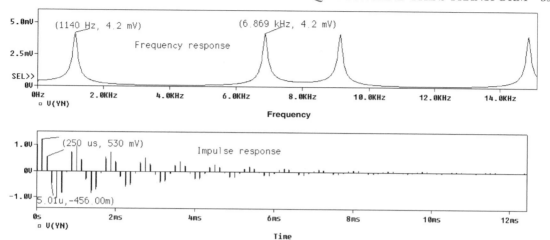

FIGURE 2.19: Impulse response.

2.12   BANDPASS IMPULSE RESPONSE

Apply a 10 us impulse to observe the impulse response shown in Fig. 2.19. Compare the graphical values to the calculated values $y(n) = x(n) + 1.2y(n-1) - 0.91y(n-2)$. Assume that $x(n)$ is an impulse whose value is 1 at zero:

$$h(0) = \delta(0) + 1.2h(0-1) - 0.91h(0-2) = 1 + 0 - 0 = \mathbf{1}$$
$$h(1) = \delta(1) + 1.2h(1-1) - 0.91h(1-2) = 0 + 1.2x1 - 0 = \mathbf{1.2}$$
$$h(2) = \delta(2) + 1.2h(n-1) - 0.91h(n-2) = 0 + 1.2x1.2 - 0.91 = \mathbf{0.53}$$
$$h(3) = \delta(3) + 1.2h(n-1) - 0.91h(n-2)\dots \text{ etc.}$$

Press keys **alt PP** to add extra plots. Copy the impulse response variable from below the x-axis (**ctrl c**) and paste the copied variable on the added plot with **ctrl v**. **Unsynchronize** the top plot (from the **Plot** menu), and click the **FFT** icon to obtain the frequency response as shown (zoom in at the beginning of the sinc response).

2.13   SIMULATING DIGITAL FILTERS USING A NETLIST

A third-order elliptical low-pass filter specification is as follows:

- Passband edge frequency = 3.2 kHz,
- Passband ripple = 0.9 dB,
- Stopband edge attenuation = 22 dB at 4.3 kHz,
- Clock frequency = 24 kHz.

The filter transfer function is

$$H(z) = \frac{(0.10285z+1)(z^2 - 0.7062z + 1)}{(z - 0.55889)(z^2 - 1.157z + 0.76494)}.$$
(2.54)

The netlist with a file extension ".cir," to implement this transfer function, is
*Must have a space or name here

```
vin 1 0 AC 1 PULSE(-1v 1v 0s 10ns 10ns 1ms 2ms);   input parameters
.ac dec 1000 1 10k;                                 The AC sweep parameters
.tran 1ms 4ms 0s 10ns;                              The Transient parameters
.func z(T) {exp(s*T)};                              z-transform
.param fs = {24kHz};                                sampling frequency
.param T = {1/fs};                                  delay
E 2 0 LAPLACE {V(1,0)}
= {0.10285*(z(T)+1)*(z(2*T)-0.70621*z(T)+1)/
+((z(T)-0.55889)*(z(2*T)
-1.1579*z(T)+0.76494))};                            Transfer function
.Probe
.end
```

PSpice models the transfer function $H(z)$ using a voltage-controlled voltage source called an **E** part. The z-transform is defined using the ".FUNC" statement, with an argument that defines the power of z (the required delay). Select **PSPICE A/D.EXE** to opens the screen as shown in Fig. 2.20.

Select **File/Open Simulation** and select **Fig2.23.cir**. This displays the netlist, so press the little sideways blue triangle to run the program and a blank screen should appear. Select the **Trace/ Add Trace** menu and enter in the **Trace Expression** box the variable **V(2)** from the list.

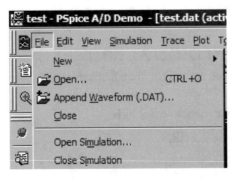

FIGURE 2.20: Select **Open Simulation**.

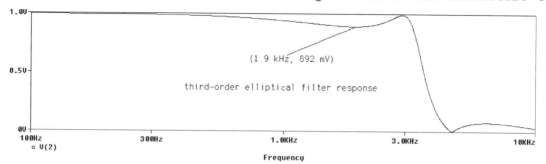

FIGURE 2.21: y-axis linear.

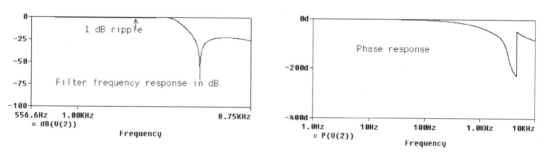

FIGURE 2.22: Frequency response in dB.

Click **OK** and the frequency response in Fig. 2.21 shows a fast roll-off rate in the transition region.

Change the y-axis to dB using the **PROBE** icon to produce the response in Fig. 2.22. Alternatively, select **dB** from the **Trace Add** right-hand list and add the variable as **dB (V(2))**.

The phase response is also plotted using **P(V(2))**. Carry out a transient analysis and observe oscillations in the output, when a square wave is applied.

2.14 DIGITAL FILTERS USING A LAPLACE PART

The **Laplace** part is a quick way to implement a delay. However, there are some penalties attached to using this part, such as speed of computation, accuracy, etc. Nevertheless, we will use it to achieve higher order filters. A first-order IIR filter transfer function is

$$H(z) = \frac{1}{1 - 0.5z^{-1}}. \qquad (2.55)$$

The delay T_s is the inverse of the sampling frequency f_s as

$$T_s = \frac{1}{f_s}. \qquad (2.56)$$

DENOM	(1-0.5*exp(-s*T))
Location X-Coordinate	140
Location Y-Coordinate	40
NUM	1
Schematics' Source Library	C:\MSimEv_8\lib\ABM.slb
Source Part	LAPLACE.Normal

FIGURE 2.23: Setting the **Laplace** part parameters.

It is a good idea to display the input and output pins of the **Laplace** part Fig. 2.24. This is necessary because you will have to flip this component when using it in IIR recursive filters (failing to do this will result in an error). The part turns green when selected, **Rclick** and select **Edit Part**. **Dlclick** the symbol and in the **User Properties** select **Pin Names Visible** and change **False** to **True**. When finished, select FILE/CLOSE and select update all. Select the part again, **Rclick** and select **Edit Properties** and enter the transfer function in the numerator (**NUM**) and denominator (**DENOM**) boxes as shown in Fig. 2.23. Replace each z^{-1} in the transfer function with $\exp(-s*T)$:

$$H(z) = \frac{1}{1 - 0.5e^{-sT}}. \tag{2.57}$$

In this example, the numerator (**NUM**) is 1 and the denominator (**DENOM**) is $1 - 0.5*\exp(-s*T)$.

The **PARAM** part in Fig. 2.24 defines the relationship between the sampling frequency f_s and the sample period T.

2.14.1 Third-Order Elliptical Filter

The netlist for a third-order elliptical filter is examined from the output file opened from the management directory and may be simulated using the following ".net" netlist, but here we

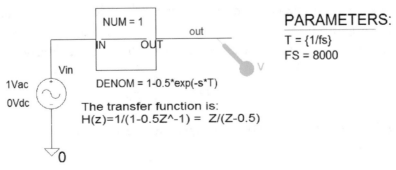

FIGURE 2.24: A **Laplace** part digital filter.

use a **Laplace** part instead:

V_Vin	SINE 0 DC 0V AC 1V
E_LAPLACE2	OUT1 0 LAPLACE {V(SINE)}
+{(1-0.60383*exp(-s*T)+0.9274*exp(-s*2*T)	
+0.10285*exp(-s*3*T))/((0.4275	
+1.412*exp(-s*T)-1.71679*exp(-s*2*T)	
+exp(-s*3*T)))}	
E_Unit_Impulse_DIFF2	IMPULSEOUT 0 VALUE
+ {V(Unit_Impulse_N00033,	Unit_Impulse_N00027)}
V_Unit_Impulse_Vstep	Unit_Impulse_N00033 0 1V
R_Unit_Impulse_delay1_RT1 0	Unit_Impulse_N00033 100
R_Unit_Impulse_delay1_RT2 0	Unit_Impulse_N00027 100
T_Unit_Impulse_delay1_T1	Unit_Impulse_N00033 0
Unit_Impulse_N00027	
+ Z0=100 TD={Impulse_width}	
.PARAM impulse_width=1u fs =8000 T={1/fs}	

The elliptical filter transfer function is

$$H(z) = \frac{1 - 0.60383z^{-1} + 0.9274z^{-2} + 0.10285z^{-3}}{-0.4275 + 1.412z^{-1} - 1.71679z^{-2} + z^{-3}}. \qquad (2.58)$$

In Fig. 2.25, **Rclick** the **Laplace** part, select **Edit Properties** to open the properties spreadsheet. The top part of the transfer function is entered in the **NUM** row, and the bottom part is entered in the **DENOM** rows.

Attach two markers from **PSpice/Marker/Advanced** menu. Set the **Analysis** tab to **Analysis type: AC Sweep/Noise, AC Sweep Type to Linear/Logarithmic, Start Frequency = 1 Hz, End Frequency = 4 kHz, Points/Decade/Decade = 1000** and simulate by pressing

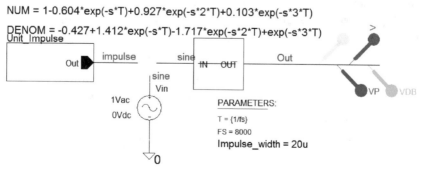

FIGURE 2.25: Elliptical filter using **ABM** part.

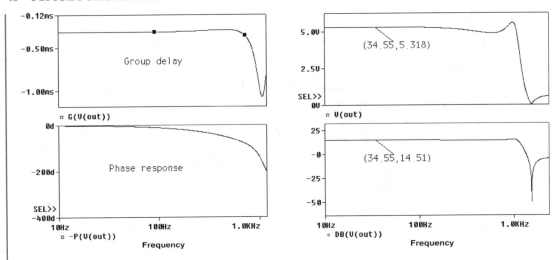

FIGURE 2.26: Elliptical filter response.

F11 to produce the filter amplitude response shown in Fig. 2.26. Included here are the amplitude response in voltage and dB, and the phase and the group delay responses.

2.14.2 Group Delay

All filters have a phase between the input and output signals (i.e., time delay between the input and output). If this phase is constant for all frequencies, then there is no problem. However, if each frequency in the input signal experiences different amounts of phase shift, i.e., the phase response is not linear, then the output signal will be distorted. One way of representing phase shift change with frequency is the concept of group delay D and is defined as the derivative of the phase shift with respect to frequency:

$$D = -\frac{d\phi}{df}. \qquad (2.59)$$

Group delay is the slope of the phase response and is important for certain classes of signals such as bandpass signals. Let us consider an amplitude-modulated carrier signal being applied to a bandpass filter. It is desirable that each frequency component in the signal has the same phase otherwise the signal will be distorted. Fig. 2.26 shows the nonlinear group delay response. A group delay marker from the **PSpice/Marker/Advanced** menu gave the group delay response, alternatively, from the **Trace Add** menu, select the **G()** operator (this was **vg()** in version 8) and substitute the variable in the brackets. Fig. 2.27 is a second-order FIR filter implemented using a **Laplace** part. The sampling frequency, fs, is entered into a **PARAM** part and hence defines the delay, since this is defined as $\{1/fs\}$. This necessitates adding rows to the **PARAM** part as discussed before. The transfer function is entered into the **NUM** part of the

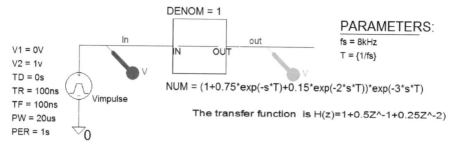

DENOM = 1

PARAMETERS:

fs = 8kHz

T = {1/fs}

V1 = 0V
V2 = 1v
TD = 0s
TR = 100ns
TF = 100ns
PW = 20us
PER = 1s

Vimpulse

NUM = (1+0.75*exp(-s*T)+0.15*exp(-2*s*T))*exp(-3*s*T)

The transfer function is H(z)=1+0.5Z^-1+0.25Z^-2)

FIGURE 2.27: Delay added to signal.

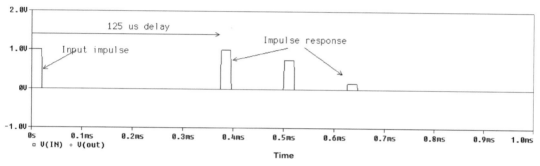

FIGURE 2.28: The input signal showing the added delay.

Laplace part, with 1 entered into the **DENOM** part. The output comprises three components $\{1, 0.75, 0.15\}$. To avoid causality problems and achieve correct levels in the plotted response, we introduce an extra 375 us delay by multiplying the transfer function by $\exp(-3*s*T)$ as shown.

The **Analysis Setup/Transient** parameters are as follows: **Run to time** = 2 ms, **Maximum step size** = 100 ns. Press **F11** to simulate and display the delayed output signal as shown in Fig. 2.28. Note that the amplitudes of the three pulses should be the same as the filter coefficients, i.e., $1, 0.75, 0.15$, where the transfer function is multiplied by three delays to fix causal problems i.e. delayed by $3*s*T$.

2.15   EXERCISES

1. In Fig. 2.9, make the coefficient negative and obtain the frequency response.

2. Repeat the cut-off frequency analysis for negative filter coefficients to express the cut-off frequency as

$$f_c = \frac{f_s}{2\pi} \cos^{-1}\left\{\frac{(a_0 + 1)^2}{4a_0}\right\}. \qquad (2.60)$$

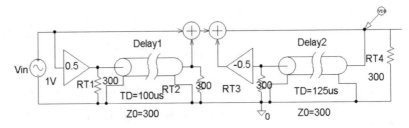

FIGURE 2.29: Bandpass digital filter.

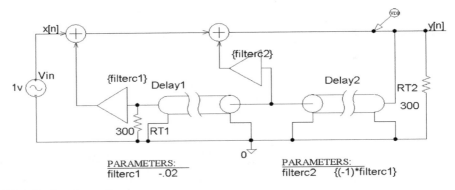

FIGURE 2.30: Bandpass filter.

3. The bandpass filter in Fig. 2.29 is formed from a cascaded low-pass and high-pass filters. The cut-off frequency for the high-pass filter is lower than the low-pass filter and means Delay1 is less than Delay2, which is not a realistic solution, but investigate anyway. Obtain the impulse and frequency responses.

4. A bandpass filter using two **PARAM** parts linked as shown in Fig. 2.30. Filterc2 coefficient is made negative by entering, in the second **PARAM** part spreadsheet a new row, **VALUE1** $= (-1) * $**filterc1** (include the curly braces). The output wire segment is shown as y[n] and is incorrect—do not use braces or spaces in wire segment names.

Obtain the frequency response for a range of filter coefficients. Make the filter coefficient range (filterc2) positive, and investigate the effect on the resonant frequency.

CHAPTER 3

Digital Convolution, Oscillators, and Windowing

3.1 DIGITAL CONVOLUTION

Digital convolution is an important concept in linear time-invariant (LTI) systems and is used to predict the output response for any input signal. The output $y(n)$ is obtained by convolving the input signal with the impulse response of the system as $y(n) = x(n)*h(n)$ (note that the asterisk (*****) represents convolution and should not be confused with multiplication as used for separating variables when entering equations in **ABM** parts). Convolution is an application of superposition, since LTI systems obey the principles of superposition, and is the sum of the product of two signals expressed as

$$y(n) = x(n)*h(n) = \sum_{k=0}^{N} x_k h(n - k). \qquad (3.1)$$

Here $x(n)$ is the input signal and $h(n)$ is the impulse response of the system. Each part of the input signal is multiplied by each part of the shifted impulse response and summed over the limits.

3.1.1 Flip and Slip Method

The schematic shown in Fig. 3.1 demonstrates digital convolution using the "flip and slip" method. The impulse response, $h(n)$, is flipped in time before being multiplied by the input signal. We could, however, also flip the input signal $x(n)$ and convolve it with $h(n)$ to produce the same result. The convolution of two signals normally starts at the origin ($n = 0$). However, it is not possible to demonstrate convolution using the zero reference point because it would mean generating noncausal flipped signals (signals before $n = 0$!!). To overcome this difficulty, we artificially create a noncausal type reference point by delaying the input signal by 750 us. If the sampling period $T = 125$ us (equal to 1/sampling frequency $= 1/8$ kHz), then multiplying the input signal by $e^{-6sT} = e^{-s750u}$ will start the input signal at 750 us. However, we will use T parts instead of LAPLACE parts and the delay is included in the input signal generator making TD in the VPULSE generator equal to 750 us.

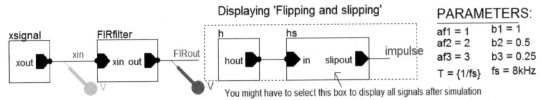

Run the log cmd file called figure3-007.cmd from Probe/File/Run Commands after simulation

FIGURE 3.1: Convolution system.

The input signal $x(n)$ is generated using three input coefficients $b_1 = 1$, $b_2 = 0.5$, $b_3 = 0.25$. These and the other coefficients are defined in a **PARAM** part for the input and impulse signals as shown in Fig. 3.2. (**Rclick/Edit Properties** to open the Param spreadsheet).

The **PARAM** part reassigns the impulse coefficients making them dependent on the FIR filter coefficients but in reverse order, so the impulse coefficients become $a_1 = a_{f3}$, $a_2 = a_{f2}$, and $a_3 = a_{f1}$. The **xsignal** box in Fig. 3.3 is generated using T parts rather than LAPLACE parts as it gives much sharper impulse signals. Note the delay TD on the input signal generator.

a1	{af3}
a2	{af2}
a3	{af1}
af1	1
af2	2
af3	3
b1	1
b2	0.5
b3	0.25
T	{1/fs}
fs	8kHz

FIGURE 3.2: Defining the filter coefficients.

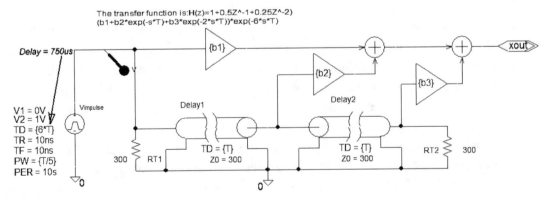

FIGURE 3.3: The input signal $x(n)$ generator.

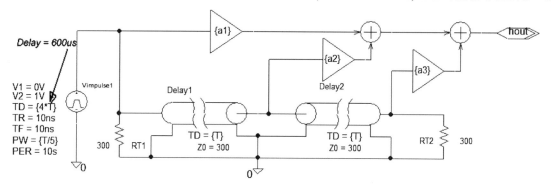

FIGURE 3.4: The impulse signal generator.

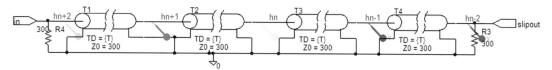

FIGURE 3.5: The slipping mechanism.

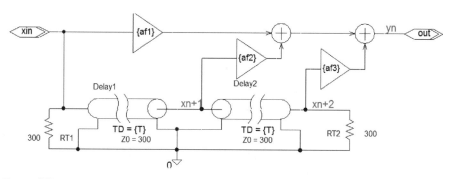

FIGURE 3.6: The convolution filter.

The flipped impulse signal $h(n)$ coefficients shown in Fig. 3.4, are also defined in the same **PARAM** part but use the curly braces {}. The impulse coefficients, a_1, a_2, a_3, are the FIR filter coefficients a_{f1}, a_{f2}, and a_{f3}, but after flipping the signal, they become a_{f3}, a_{f2}, and a_{f1}.

The impulse signal is "flipped and slipped" and then passed by the input signal where it is multiplied by the input signal and the product summed. The slipping is achieved using the schematic block called **h** and **hs** in Fig. 3.5, with the line segment named as shown. Each block contains a transmission line with delay T (defined in the **Param** part as $\{1/f_s\}$, where f_s is the sampling frequency).

The block **FIRfilter** in Fig. 3.6 is the main convolution schematic with two delays and three filter coefficients whose values are defined in the **PARAM** part as $\{a_{f1}\}$, $\{a_{f2}\}$, and $\{a_{f3}\}$.

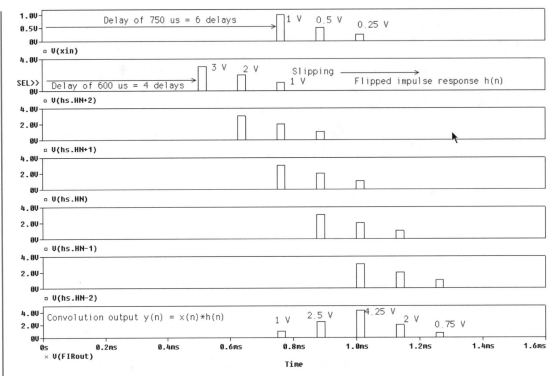

FIGURE 3.7: "Flip and slip" technique.

The convolution process is shown in Fig. 3.7. The zero reference is now located at 750 us, i.e., equal to six delay units = 6 × 125 us. A running sum of products is performed from this reference point. The transient parameters are as follows: **Run to time** = 2 ms. Tick **Skip initial transient** in the **Transient Setup** menu, and press the **F11** key to simulate.

From the **PROBE** screen, select **File/Run Commands . . .** and load the log file **Figure 3-6.cmd**. This is a text file of the commands to separate the signals in the right order that was originally created by selecting **File/Log Commands** to record all keystrokes pressed. This option must be unticked from Probe/File menu when finished recording. Note: After simulation, it seems necessary to select the **hs** block (**shift + D**) to display the delayed signals in that block (**shift + A** gets you back to the main schematic). The 'product and sum' figures shown in Table 3.1 correspond to the values displayed in the bottom pane in Fig. 3.7.

Convolution of signals $x(n)$ and $h(n)$ in the time domain is equivalent to multiplication in the z-domain, i.e.,

$$y(n) = x(n)*h(n). \tag{3.2}$$

TABLE 3.1: Convolution in table format

$y(n)$	$x(n)h(n)$			$\sum xh$
$y(1)$	$h_0\,x_0 = 1 \times 1$			1
$y(2)$	$h_0\,x_1 = 1 \times 0.5$	$h_1\,x_0 = 2 \times 1$		2.5
$y(3)$	$h_0\,x_2 = 1 \times 0.25$	$h_1\,x_1 = 2 \times 0.5$	$h_2\,x_0 = 3 \times 1$	4.25
$y(4)$		$h_1\,x_2 = 2 \times 0.25$	$h_2\,x_1 = 3 \times 0.5$	2
$y(5)$			$h_2\,x_2 = 3 \times 0.25$	0.75

The asterisk denotes convolution not multiplication:

$$Y(z) = X(z) x\, H(z). \tag{3.3}$$

Transforming input and impulse signals and multiplying together yields the same result as convolving the signals:

$$X(z) = 1 + 0.5z^{-1} + 0.25z^{-2}$$
$$H(z) = 1 + 2z^{-1} + 3z^{-2}$$

$$
\begin{aligned}
&= 1 + 0.5z^{-1} + 0.25z^{-2} \\
&\quad\ \ 2z^{-1} \ \ + 1z^{-2} \ \ \ \ + 0.5z^{-3} \\
&\quad\ \ \ \ \ \ \ \ \ + 3z^{-2} \ \ \ \ + 1.5z^{-3} + 0.75z^{-4}
\end{aligned}
$$

$$Y(z) = 1 + 2.5z^{-1} + 4.25z^{-2} + 2z^{-3} + 0.75z^{-4}$$

The coefficients in bold are the same as those from the flip and slip convolution method investigated previously. To show that this is true, create the schematic in Fig. 3.8. The two transfer functions are multiplied using two **Laplace ABM** parts, with each function $X(z)$ and $H(z)$ entered into **NUM** box (the numerator). The bitmap picture of the convolution output was created by selecting **Probe/Window/Copy to Clipboard**. Save this as a bitmap file using Paint Shop Pro or similar, and import into the schematic from the schematic menu **Place/Picture**.

Of course, we could replace both blocks within a single **Laplace** part whose transfer function is

$$(1 + 0.5^* \exp(-s^*T) + 0.25^* \exp(-s^*2^*T))^*(1 + 2^* \exp(-s^*T)$$
$$+ 3^* \exp(-s^*2^*T))^* \exp(-s^*4^*T). \tag{3.4}$$

Carry out a transient analysis and verify that the output is the same as before.

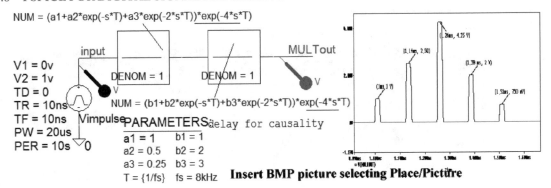

NUM = (a1+a2*exp(-s*T)+a3*exp(-2*s*T))*exp(-4*s*T)

input   MULTout

V1 = 0v
V2 = 1v
TD = 0
TR = 10ns
TF = 10ns   Vimpulse
PW = 20us
PER = 10s

DENOM = 1   DENOM = 1

V

NUM = (b1+b2*exp(-s*T)+b3*exp(-2*s*T))*exp(-4*s*T)

PARAMETERS delay for causality

a1 = 1   b1 = 1
a2 = 0.5   b2 = 2
a3 = 0.25   b3 = 3
T = {1/fs} fs = 8kHz   **Insert BMP picture selecting Place/Picture**

FIGURE 3.8: Multiplication in the z-domain.

3.2   DSP SINUSOIDAL OSCILLATOR

We design a DSP oscillator by taking the z-transform of the output—a cosine signal $\cos(n\theta)$.

From the z-tables in Appendix B, we see that the z-transform for $\cos n\theta u(n)$ is $\frac{z(z-\cos\theta)}{z^2-2z\cos\theta+1}$. We obtain a difference equation using this result so that we can draw a block diagram to implement this oscillator. The transfer function is

$$H(z) = \frac{Y(z)}{X(z)} = \frac{z^2 - z\cos\theta}{z^2 - 2z\cos\theta + 1} \div z^2 = \frac{1 - \cos\theta z^{-1}}{1 - 2z^{-1}\cos\theta + z^{-2}}. \qquad (3.5)$$

Manipulate (3.5) and write the difference equation in the time domain as

$$y(n) = x(n) - \cos\theta x(n-1) + 2\cos\theta y(n-1) - y(n-2). \qquad (3.6)$$

Since an oscillator has no external input, we set $x(n)$ to zero, hence the difference equation is

$$y(n) = 2\cos\theta y(n-1) - y(n-2). \qquad (3.7)$$

Fig. 3.9 shows an impulse signal lasting for a very short time only in order to "kick-start" the oscillator into producing sustained oscillations but starting at a known value. The initial values for the oscillator are $y(-2) = -A\sin\theta$ and $y(-1) = 0$. z-transforming (3.7) yields

$$Y(z) = 2\cos\theta Y(z)z^{-1} - Y(z)z^{-2}. \qquad (3.8)$$

Rearranging (3.8) yields

$$Y(z) - 2\cos\theta Y(z)z^{-1} + Y(z)z^{-2} = 0 \Rightarrow Y(z)[1 - 2\cos\theta z^{-1} + z^{-2}] = 0. \qquad (3.9)$$

A general form for (3.9) is

$$Y(z)(1 - b_1 z^{-1} + b_2 z^{-2}) = 0. \qquad (3.10)$$

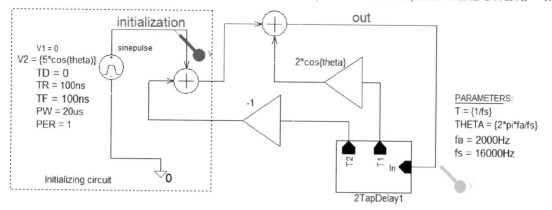

FIGURE 3.9: Digital oscillator.

The output $Y(z)$ cannot be zero if sustained oscillations are required at the output, therefore $(1 - b_1 z^{-1} + b_2 z^{-2}) = 0$. To obtain the coefficient values to achieve this condition, we must first get expressions for the roots of (3.10) using $-b/2a \pm \sqrt{b^2 - 4ac}/2a$, where $a = 1$, $b = b_1$, and $c = b_2$:

$$\frac{b_1}{2} \pm \frac{\sqrt{b_1^2 - 4b_2}}{2} = \frac{b_1}{2} \pm \frac{\sqrt{-1}\sqrt{4b_2 - b_1^2}}{2} = \frac{b_1}{2} \pm j\frac{\sqrt{4b_2 - b_1^2}}{2}. \qquad (3.11)$$

Second-order systems have complex conjugate poles, hence $4b_2$ must be greater than b_1, so we end up with the square root of a negative number but this is fixed by rearranging the square root and introducing the imaginary number $j = \sqrt{(-1)}$. To produce sustained oscillation in a system, the poles must be located on the unit circle and the system is then said to be marginally stabile. Fig. 3.10 is a plot of the pole–zeros where the poles are located on the unit circle with an angle of $\pi/2$ between them, or $45°$ to the real axis. Since the radius of the z-plane is unity, then the horizontal component is written as

$$b_1/2 = \cos \theta \Rightarrow b_1 = 2 \cos \theta. \qquad (3.12)$$

The imaginary part of the pole (the vertical component) is

$$\frac{\sqrt{4b_2 - b_1^2}}{2} = \sin \theta. \qquad (3.13)$$

Substituting for b_1 from (3.12) into (3.13) and squaring yields

$$4b_2 - (2 \cos \theta_0)^2 = 4 \sin^2 \theta_0 \qquad (3.14)$$

$$4b_2 = 4 \sin^2 \theta + 4 \cos^2 \theta = 4(\sin^2 \theta + \cos^2 \theta) \Rightarrow b_2 = 1. \qquad (3.15)$$

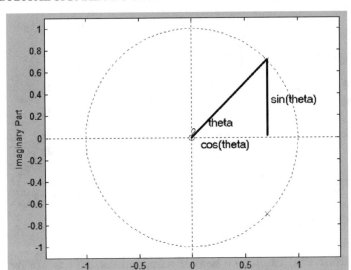

FIGURE 3.10: Pole–zero plot.

An expression for digital frequency is $\theta = 2\pi f_a / f_s$, and is obtained from (3.15) as

$$\cos \theta = b_1/2 \Rightarrow \theta = \cos^{-1}(b_1/2) = 2\pi f_a/f_s. \qquad (3.16)$$

The sampling frequency is 16 kHz for a sustained frequency of 2 kHz:

$$f_a = \cos^{-1}\left(\frac{b_1}{2}\right) f_s/2\pi = \cos^{-1}(0.707)16000/2\pi = 2000. \qquad (3.17)$$

For the values used in the simulation, $b_1 = 2 \cos \theta = 2 \cos(0.7854) = 1.4142$ and $b_2 = -1$:

$$H(z) = \frac{1}{1 - 1.4142z^{-1} + z^{-2}}. \qquad (3.18)$$

Using Matlab, we may plot in Fig. 3.10 the pole–zero using the mfile **zplane(b,a)**, where $a = (1, -1.414, 1)$ and $b = 1$. This is a case of margin stability where the poles are located on the unit circle.

Output File Options/Print values in the output file = 100 ns, **Run to time** = 5 ms, and **Maximum step size** = 10 ns. Press the **F11** key to simulate. The decaying kick-start and output signals are shown in Fig. 3.11. Select **Plot/Unsynchronize** to unsynchronize the top plot and change the x-axis range to see the very narrow impulse.

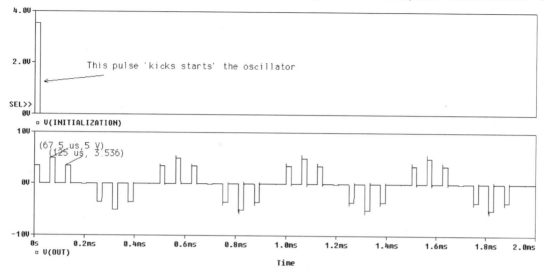

FIGURE 3.11: The digital output and the initialization signals.

3.3   EXERCISES

1. Investigate the digital state variable filter shown in Fig. 3.12 [ref: 10 Appendix A].

2. The response for each output is shown in Fig. 3.13. Investigate changing the Q and f_c controls.

3. The state variable filter can be modified for quadrature sin/cosine production as shown in Fig. 3.14.
    The outputs are shown in Fig. 3.15.

4. Investigate correlation using the convolution schematic examined previously.
    Correlation differs from convolution because the impulse (or input signal) is not "flipped" and examines the similarity between signals.

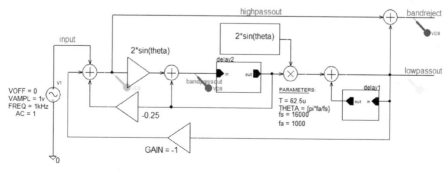

FIGURE 3.12: State variable digital filter.

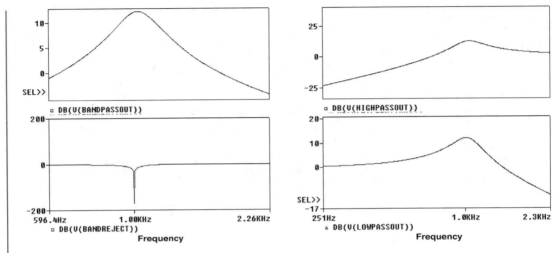

FIGURE 3.13: State variable filter responses.

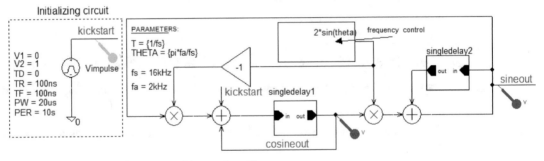

FIGURE 3.14: Digital state variable quad oscillator.

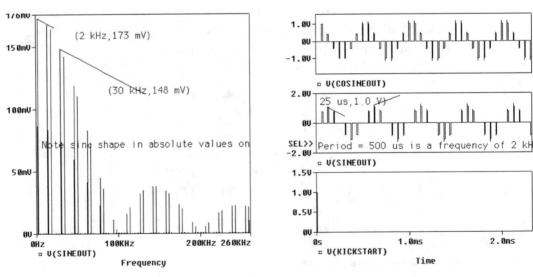

FIGURE 3.15: 2 kHz cosine and sine output.

CHAPTER 4

Digital Filter Design Methods

4.1 FILTER TYPES

Filters are used to modify the frequency spectrum of signals and are available in five basic filter types:

- Low-pass: it passes low frequencies but attenuates high frequencies.
- High-pass: it passes high frequencies but attenuates low frequencies.
- Bandpass: it passes a band of frequencies and attenuates frequencies outside that band.
- Bandstop: it stops a band of frequencies but passes all other frequencies.
- All-pass: it is used to modify the phase response.

We examined analog passive and active filters in Book2. [Ref: 1 Appendix A], but in this chapter we examine a range of digital filters only. A *recursive* IIR filter output depends on *past* output values, whereas a nonrecursive *FIR* filter has an output that depends only on the present input signal. Digital filters are classified from the impulse response into two types: a finite impulse response (FIR) filter has a finite number of terms in the response and is nonrecursive, whereas an infinite impulse response (IIR) filter has an infinite number of terms in the response and is called a recursive filter (in practice, we cannot have an infinite number in the response). A difference equation relates the input $x(n)$, output $y(n)$, and filter coefficients a_k and b_k as

$$y(n) = \sum_{k=0}^{M} a_k x(n-k) - \sum_{k=1}^{L} b_k y(n-k). \qquad (4.1)$$

An IIR filter meets a specification using a lower filter order when compared to the design using a FIR filter. This gives greater computational efficiency, and the phase delay and data storage requirements are also much lower when compared to the requirements of an FIR filter. A disadvantage with recursive filters, however, is that they have a *nonlinear phase response*, and also greater coefficient sensitivity. A nonlinear phase results in signal spectral components being delayed by different amounts and this may present problems in certain applications, such as with passband signals, or video and audio signals. (Group delay was discussed in a previous chapter as a means of investigating filter delay problems.) Coefficient sensitivity arises when quantization errors cause filter coefficients to be rounded off and change the desired frequency response (it could also make the filter unstable). Expressing (4.1) in the z-domain yields

$$Y(z) = \sum_{k=0}^{M} a_k X(z) z^{-k} - \sum_{k=1}^{L} b_k Y(z) z^{-k}. \qquad (4.2)$$

Thus, the transfer function is

$$H(z) = \frac{Y(z)}{X(z)} = \frac{\displaystyle\sum_{k=0}^{M} a_k z^{-k}}{1 + \displaystyle\sum_{k=1}^{L} b_k z^{-k}}. \qquad (4.3)$$

Butterworth, Chebychev (types I and II), and Elliptic IIR filter types use a mapping technique to convert low-pass analog filter designs to an equivalent digital filter. Elliptic filters have equiripple in the passband and stopband regions but have the steepest roll-off in the transition region when compared to other filter types of the same order. These filters also have minimum transition widths but the worst coefficient sensitivity and a nonlinear phase response characteristic. We now investigate different filter configurations and see how we can reduce the number of delay elements in the final design. This saves on the number of computations and hence, the overall speed of operation is increased.

4.2   DIRECT FORM 1 FILTER

The two parts of (4.2) define FIR and IIR filters. When $M = L$, we have a canonic form realization (a digital filter is said to be *canonic* when it uses a minimum number of delays to produce a desired frequency response). The second part defines recursive filter with the output fed back to the input, whilst the first part defines FIR filter types with feed-forward components only. We can rewrite (4.3) for a Direct Form 1 filter transfer function:

$$H(z) = \sum_{k=0}^{M} a_k z^{-k} \frac{1}{1 + \displaystyle\sum_{k=1}^{L} b_k z^{-k}}. \qquad (4.4)$$

Here the filter zeros are followed by the poles. The recursive Butterworth second-order high-pass filter in Fig. 4.1 has a cut-off frequency = 6.38 kHz and sampling frequency of

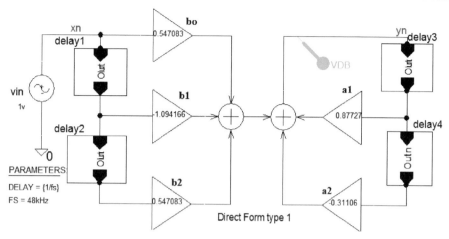

FIGURE 4.1: Direct form type 1 (zeros–poles).

48 kHz. The transfer function for type 1 filter is

$$H(z) = \frac{Y(z)}{X(z)} = \frac{b_0 + b_1 z^{-1} + b_2 z^{-2}}{1 - a_1 z^{-1} - a_2 z^{-2}} = (b_0 + b_1 z^{-1} + b_2 z^{-2})\left(\frac{1}{1 - a_1 z^{-1} - a_2 z^{-2}}\right).$$

(4.5)

The filter coefficients are as follows: $b_0 = 0.547083$, $b_1 = -1.094166$, $b_2 = 0.547083$, $a_1 = -0.087727$, $a_2 = 0.31106$.

4.3 DIRECT FORM 2 FILTER

This differs from Direct Form 1 since the poles are followed by the zeros. We also have a reduced number of delays, which is good. We derive the transfer function by calling the output from the first three-input summer **P**. Thus, from the output in Fig. 4.2

$$P = a_1 z^{-1} P + a_2 z^{-2} P + X(z) \Rightarrow P(1 - a_1 z^{-1} - a_2 z^{-2}) = X(z) \qquad (4.6)$$

and

$$Y(z) = P(b_0 + b_1 z^{-1} + b_2 z^{-2}). \qquad (4.7)$$

Dividing (4.7) by (4.6) produces the transfer function for the pole–zero Direct Form 2 filter:

$$H(z) = \frac{Y(z)}{X(z)} = \frac{1}{1 + \sum_{k=1}^{L} b_k z^{-k}} \sum_{k=0}^{M} a_k z^{-k} = \left(\frac{1}{1 - a_1 z^{-1} - a_2 z^{-2}}\right)(b_0 + b_1 z^{-1} + b_2 z^{-2}).$$

(4.8)

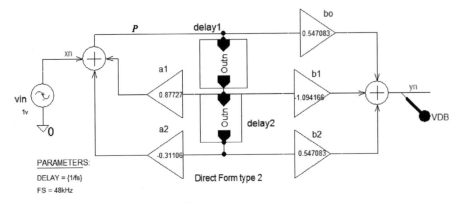

FIGURE 4.2: Direct Form 2 (poles–zeros).

This is still a recursive Butterworth second-order high-pass filter where the sample frequency is 48 kHz and the cut-off frequency is 6.38 kHz. The filter coefficients are as follows: $b_0 = 0.547083$, $b_1 = -1.094166$, $b_2 = 0.547083$, $a_1 = -0.087727$, $a_2 = 0.31106$.

4.4   THE TRANSPOSE FILTER

The transpose filter configuration is shown in Fig. 4.3. We can see that the number of delays is the same as DF type 2. The transfer function for the type 2 filter is derived by considering the R and Q outputs in that sequence. The R output is therefore

$$R = a_2 Y(z) + b_2 X(z). \tag{4.9}$$

The Q output is obtained by substituting (4.9) as

$$Q = a_1 Y(z) + b_1 X(z) + Rz^{-1} \Rightarrow Q = a_1 Y(z) + b_1 X(z) + a_2 Y(z)z^{-1} + b_2 X(z)z^{-1}. \tag{4.10}$$

The output (P) is the final output $Y(z)$:

$$Y(z) = Qz^{-1} + b_0 X(z). \tag{4.11}$$

Substituting (4.1), hence, we can write

$$Y(z) = Qz^{-1} + b_0 X(z) \Rightarrow Y(z)$$
$$= a_1 Y(z)z^{-1} + b_1 X(z)z^{-1} + a_2 Y(z)z^{-2} + b_2 X(z)z^{-2} + b_0 X(z) \tag{4.12}$$

$$Y(z)[1 - a_1 z^{-1} - a_2 z^{-2}] = X(z)[b_0 + b_1 z^{-1} + b_2 z^{-2}] \tag{4.13}$$

$$H(z) = \frac{Y(z)}{X(z)} = \frac{b_0 + b_1 z^{-1} + b_2 z^{-2}}{1 - a_1 z^{-1} - a_2 z^{-2}}. \tag{4.14}$$

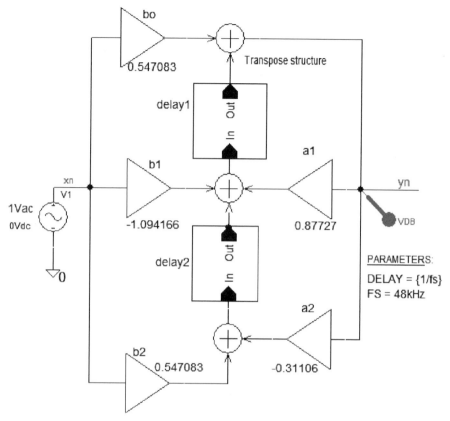

FIGURE 4.3: Transpose filter type.

Set the Analysis tab to Analysis type: AC Sweep/Noise, AC Sweep Type to **Linear/Logarithmic, Start Frequency** = 100, **End Frequency** = 100k, **Total Points/ Decade** = 1000. Press **F11** to simulate. The response is the same for all three filter types and is shown in Fig. 4.4.

4.5   CASCADE AND PARALLEL FILTER REALIZATIONS

To produce high-order filter, it is easier to design first and second-order filter and then connect them in series. This arrangement is called a cascade filter and is shown in Fig. 4.5. Shown in the same diagram is a parallel arrangement.

4.5.1   Digital Filter Specification

Equiripple digital filter specification includes the filter order N, passband ripple R_p, and stopband ripple R_s. The ripple is quoted in dB, or as a factor. Consider the following FIR low-pass filter specification:

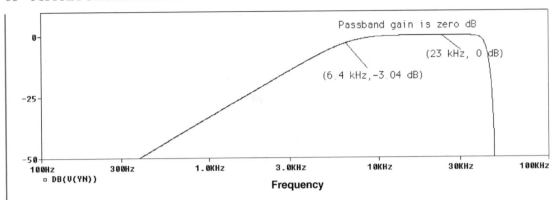

Passband gain is zero dB

(23 kHz, 0 dB)

(6.4 kHz,-3.04 dB)

□ DB(V(YN))

Frequency

FIGURE 4.4: High-pass filter for types 1 and 2.

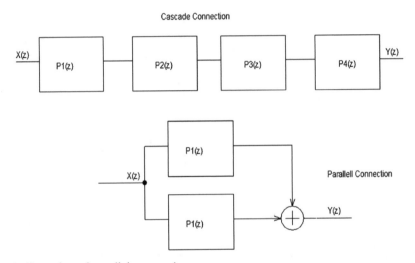

FIGURE 4.5: Cascade and parallel connection.

- Sampling frequency $f_T = 8$ kHz.
- Passband edge frequency $f_p = 1$ kHz.
- Stopband edge frequency $f_s = 1.5$ kHz.
- Maximum passband ripple $\alpha_p = 5$ dB, (or passband ripple factor $\delta_p = 0.4377$).
- Minimum stopband ripple $\alpha_s = 40$ dB, (or stopband ripple factor $\delta_s = 0.01$).

The maximum passband ripple is $1 + \delta_p$ and the minimum passband ripple is $1 - \delta_p$, hence the peak-to-peak passband ripple $2\alpha_p$ is $2(10 \log((1 + \delta_P))/(1 - \delta_P))$ dB, or the peak ripple α_p is

$$\alpha_p = 10 \log_{10}[(1 + \delta_p)/(1 - \delta_p)] \text{ dB.} \qquad (4.15)$$

Alternatively, the peak ripple is calculated as

$$\alpha_p = 20\log_{10}(1 - \delta_p) \Rightarrow \delta_p = 1 - 10^{(-\alpha_p/20)} = 1 - 10^{(-5/20)} = 0.4377. \qquad (4.16)$$

The minimum stopband attenuation is

$$\alpha_s = -20\log_{10}(\delta_s) \text{ dB} \Rightarrow \delta_s = 10^{(-\alpha_s/20)} = 10^{(-40/20)} = 0.01. \qquad (4.17)$$

The FIR filter length is calculated as

$$L = 1 - \frac{10\log(\delta_p \delta_s) - 13}{14.6\frac{f_s - f_p}{f_T}} = 1 - \frac{10\log(0.4377 \times 0.01) - 13}{14.6\frac{500}{8000}} = 12. \qquad (4.18)$$

The filter order is 1 less than the filter length, i.e., 11. The maximum passband attenuation is defined as

$$\alpha_{\max} \cong -20\log_{10}(1 - 2\delta_p) \text{ dB} \cong 2\alpha_p \qquad (4.19)$$

This approximation is valid since $\delta_p \ll 1$ in most cases. Fig. 4.6 shows the passband ripple δ_p and the stopband ripple δ_s. The transition bandwidth is f_T, and the sampling frequency is F_T [ref 9: Appendix A].

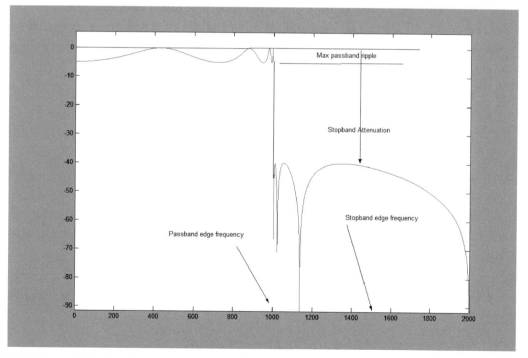

FIGURE 4.6: Elliptical filter amplitude response.

The Matlab mfile **[b, a] = ellip(N, Rp, Rs, Wn)** returns the filter coefficients for an Nth order low-pass digital elliptic filter with R_p decibels of ripple in the passband, and in the stopband R_s decibels. The cut-off frequency W_n must be $0 < W_n < 1$, with 1 corresponding to half the sample rate. Plot with **[h,f] = freqz(b,a,4000,8000)** and plot(f,20*log10(abs(h))). In this example, $N = 12$, $R_p = 0.4377$, $R_s = 0.01$, $W_n = 0.25$. Thus **[b, a] = ellip(12, 5, 40, 0.25)**, where **b** is the numerator coefficient matrix and **a** is the denominator matrix coefficients, returns

$$\mathbf{b} = 0.0184, -0.1264, 0.458, -1.1099, 1.9944, -2.7873, 3.1093,$$
$$-2.7873, 1.9944, -1.1099, 0.4580, -0.1264, 0.0184$$
$$\mathbf{a} = 1, -8.8289, 38.1653, -105.639, 207.391, -303.216, 337.908,$$
$$-288.979, 188.258, -91.215, 31.276, -6.842, 0.7281.$$

It is left as an exercise to implement this in PSpice using the **Laplace** parts for the delays and coefficients.

4.6   THE BILINEAR TRANSFORM

A digital filter can be designed using the bilinear transform to convert an analog filter prototype in the s-domain, to the z-domain. The transform maps the left-hand side of the s-plane to the inside of the unit circle in the z-plane resulting in a stable filter configuration. However, the mapping is not perfect at frequencies close to half the sampling frequency and so we must prewarp the analog signal using a tan function in order to wrap the left-hand side of the s-plane onto the inside of the unit circle z-plane. From the z-transform we write

$$z = e^{sT} = \frac{e^{sT/2}}{e^{-sT/2}} \approx \frac{1 + sT/2}{1 - sT/2}. \qquad (4.20)$$

Crossing multiplying and rearranging in terms of s yields

$$s = \frac{2}{T}\frac{1 - z^{-1}}{1 + z^{-1}} = \frac{2}{T}\frac{z - 1}{z + 1}. \qquad (4.21)$$

Note that this has a similar form to the transmission line equation for the reflection coefficient on which the Smith chart is based. We may rework (4.21) to obtain a warping function:

$$H_d(z) = H_a(e^{j\omega_d T}) = \frac{2}{T}\frac{z - 1}{z + 1}\bigg|_{z = e^{j\omega_d T}} = \frac{2}{T}\frac{e^{j\omega_d T} - 1}{e^{j\omega_d T} + 1} = \frac{2}{T}\frac{e^{j\omega_d T/2}(e^{j\omega_d T/2} - e^{-j\omega_d T/2})}{e^{j\omega_d T/2}\left(e^{j\omega_d T/2} + e^{-j\omega_d T/2}\right)} \qquad (4.22)$$

$$H_d(z) = j\frac{2}{T}\frac{(e^{j\omega T/2} - e^{-j\omega T/2})/2j}{(e^{j\omega T/2} + e^{-j\omega T/2})2} = j\frac{2}{T}\frac{\sin(j\omega T/2)}{\cos j(\omega T/2)} = j\frac{2}{T}\tan\left(\frac{\omega T}{2}\right) \qquad (4.23)$$

$$H_d(z) = H_a(j\omega_a) = j\frac{2}{T}\tan\left(\frac{\omega_d T}{2}\right) \Rightarrow \qquad (4.24)$$

$$\omega_a = 2\pi f_d = \frac{2}{T}\tan\left(\frac{\omega_d T}{2}\right) \Rightarrow f_a = \frac{1}{\pi T}\tan(\pi f_d T). \qquad (4.25)$$

We can express the digital cut-off frequency in terms of the analog cut-off frequency using the arctan function.

$$f_d = \frac{1}{\pi T}a\tan(\pi f_a T). \qquad (4.26)$$

4.6.1 Designing Digital Filters Using the Bilinear Transform Method

An analog low-pass filter transfer function is required to meet a specification using approximation loss function analysis. The infinitely long line of the imaginary axis in the s-plane is mapped to a finite interval between 0 and π in the z-plane. It is desirable that the amplitude response is the same in both domains. However, equal increments along the unit circle in the z-plane correspond to larger and larger bandwidths along the imaginary axis in the s-plane. We must introduce frequency warping because the analog frequency axis is infinite, whilst the digital frequency axis is finite. Given an analog cut-off frequency, we need to obtain the same cut-off frequency in the digital filter, or vice versa. The digital cut-off frequency is warped using $f_a = 1/(\pi T)\tan(\pi f_d T)$ where the angle is in radians. This generates an equivalent analog frequency, which is then applied to the loss function to denormalize it for a particular specification.

Apply the bilinear transform to the denormalized inverted loss function (the analog transfer function, $H(s)$), to get the required z-domain transfer function. To illustrate this technique, we select a first-order low-pass Butterworth approximation loss function $A(\$) = \$ + 1$. Inverting and denormalizing this function produces a first-order analog low-pass filter with a cut-off frequency, ω_c:

$$A(\$) = \$ + 1 \Rightarrow H_a(\$) = \left.\frac{1}{s+1}\right|_{\$=s/\omega_c} = \frac{1}{s/\omega_c + 1} = H(s) = \frac{\omega_c}{s + \omega_c}. \qquad (4.27)$$

To illustrate this technique, assume a 1 kHz digital cut-off frequency and a sampling period $T = 1/f_s = 1/8000 = 125$ us.

$$f_a = \frac{1}{\pi T}\tan(\pi f_d T) = \frac{1}{\pi 125u}\tan(\pi 1000 \times 125\ u) = 1054\ \text{Hz}. \qquad (4.28)$$

The **AC Sweep** parameters are as follows: **Points/Decade** = 1001, **Start Frequency** = 10 Hz, **End Frequency** = 4 kHz. Press **F11** to simulate. Equation (4.28) thus prewarps the digital cut-off frequency. We can see the difference between the analog cut-off frequency

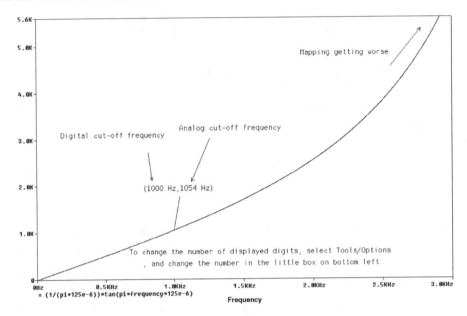

FIGURE 4.7: Input signal frequency prewarping.

and the prewarped digital cut-off frequency in Fig. 4.7 by selecting **Trace Add,** and entering (1/(pi*125e-6))*tan(pi*frequency*125e-6) in the **Trace Expression** box.

The 1 kHz digital cut-off frequency is now warped to a cut-off analog frequency of 1054 Hz. Poor warping results if the cut-off frequency is too close to the sampling frequency. The analog transfer function after prewarping is

$$H_a\,(\$) = \left.\frac{1}{\$+1}\right|_{\$=s/\omega_c} = H_a\,(s) = \frac{1}{s/\omega_c+1} = \frac{\omega_c}{s+\omega_c} = \frac{2\pi\,1054}{s+2\pi\,1054}. \qquad (4.29)$$

To obtain an equivalent digital filter to this first-order analog filter, we apply the bilinear transform to (4.29):

$$H_d(z) = \left.H_a(s)\right|_{s=\left(\frac{2}{T}\right)\frac{z-1}{z+1}} \qquad (4.30)$$

$$H(z) = \frac{2\pi\,1054}{2/\,T(1-z^{-1})/(1+z^{-1})+2\pi\,1054} = \frac{\pi\,1054T\left(1+z^{-1}\right)}{\left(1-z^{-1}\right)+\pi\,1054T(1+z^{-1})}$$

$$= \frac{0.4139\left(1+z^{-1}\right)}{1.4139-0.5861z^{-1}} = \frac{0.2927\left(1+z^{-1}\right)}{1-0.4145z^{-1}} \qquad (4.31)$$

$$H(z) = \frac{Y(z)}{X(z)} = \frac{0.2927\left(1+z^{-1}\right)}{1-0.4145z^{-1}} \Rightarrow Y(z)(1-0.4145z^{-1}) = X(z)0.2927(1+z^{-1}). \quad (4.32)$$

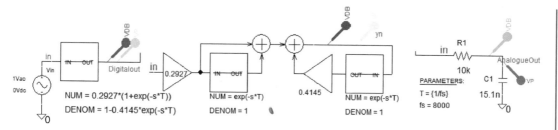

FIGURE 4.8: Analog CR LPF and its digital equivalent.

The difference equation (DE) is, therefore,

$$y(n) = 0.2927x(n) + 0.2927x(n-1) + 0.4145y(n-1). \qquad (4.33)$$

From the DE, draw the first two filters shown in Fig. 4.8. The first filter uses a **Laplace** part where the equation in (4.31) is entered in the **NUM**erator and **DENOM**inator of the part. The second filter uses the same part to achieve the delays required by (4.33). The analog filter capacitance, $C1$, is calculated, using the -3 dB cut-off expression with $R1 = 10\,k\Omega$, as

$$C = \frac{1}{2\pi f_c R} = \frac{1}{2\pi\,1054 \times 10^4} = 15.1 \text{ nF.} \qquad (4.34)$$

The **AC Sweep** parameters are as follows: **Points/Decade** $= 1001$, **Start Frequency** $= 10$ Hz, **End Frequency** $= 4$ kHz. Press **F11** to simulate to produce the amplitude and phase response for the three filters as shown in Fig. 4.9. From the **PROBE File** menu tick **Run Commands** and select **Fig.4-005.cmd** to separate the variables and plot the correct display. The **ABM** and digital filter responses will be superimposed on each other if the calculations are correct. The analog response deviates at frequencies close to half the sampling frequency and so the choice of sampling frequency affects the mapping and should be chosen much higher than that required by the Nyquist rate. For example, making the digital cut-off frequency a tenth of the sampling frequency results in good mapping to the analog cut-off frequency response. However, investigate the case for $f_c = f_s/4$ to show how poor mapping results. We may apply this transform to design bandpass and bandstop digital filters also.

When designing bandpass filters we use the bandpass frequency transformation:

$$\$ = \frac{s^2 + \omega_0^2}{Bs}. \qquad (4.35)$$

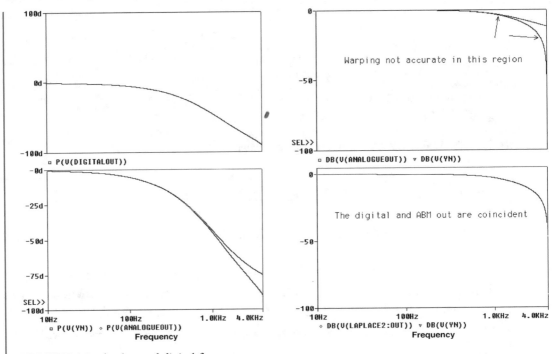

FIGURE 4.9: Analog and digital frequency response.

The resonant frequency is $f_0 = \sqrt{f_{p1} f_{p2}}$ Hz, and the -3 dB bandwidth is $B = \omega_{p2} - \omega_{p1}$ r s^{-1} [ref: 8 Appendix A]. However, we must warp the upper and lower cut-off frequencies to generate the correct digital bandwidth and center frequency using the following expressions:

$$BW = f_H - f_L = \frac{1}{\pi T}[\tan \pi f_{DH} T - \tan \pi f_{DL} T] \qquad (4.36)$$

$$f_C^2 = f_H f_L = \frac{1}{(\pi T)^2} \tan \pi f_{DH} T \tan \pi f_{DL} T. \qquad (4.37)$$

Substitute values into these two expressions and then substitute them back into (4.35). This is then substituted into the selected inverted loss function to generate a digital bandpass transfer function. Similarly, to generate a bandstop digital filter we use the same procedure but the transformation formula (4.35) must be inverted.

4.7   THE IMPULSE-INVARIANT FILTER DESIGN TECHNIQUE

The sequence of events for this method is to obtain an analog transfer function and get its impulse response, $h(t)$. This response is then sampled to produce a sampled impulse response, $h(nT)$. We then z-transform this function and from this we can write the digital transfer function, $H(z)$. The sampled impulse response should be of the same shape as the analog impulse response, hence the name invariant. The impulse-invariant digital filter design technique is not as popular, or as useful, for designing digital filters as other methods but, nevertheless, it has its uses. The digital filter obtained using this method produces an output that is an approximation for all inputs other than an impulse signal where it is exactly right. Here, we illustrate the technique using first- and second-order filter circuits. Consider a first-order analog low-pass filter whose transfer function is

$$H(s) = \frac{V_0(s)}{V_{in}(s)} = \frac{\omega_c}{s + \omega_c} \Rightarrow V_0(s) = V_{in}(s)\frac{\omega_c}{s + \omega_c}. \qquad (4.38)$$

The output voltage for a unit impulse is

$$V_0(s) = \delta(0)\frac{\omega_c}{s + \omega_c} \Rightarrow \omega_c\frac{1}{s + \omega_c}. \qquad (4.39)$$

The impulse response, $h(t)$, is then obtained by getting the inverse Laplace transform of (4.39) (use the Laplace tables included in the Appendix):

$$h(t) = \omega_c\, e^{-\omega_c T}. \qquad (4.40)$$

4.7.1   Impulse Function Generation

Fig. 4.10 shows how an impulse is generated by applying a step function to a differentiator.
    Differentiating a step signal yields the impulse signal response shown in Fig. 4.11.

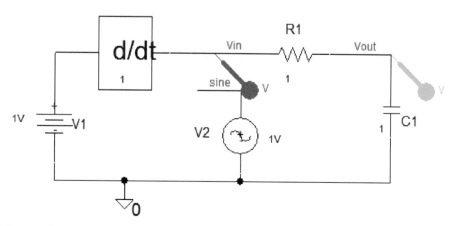

FIGURE 4.10: Producing an impulse function.

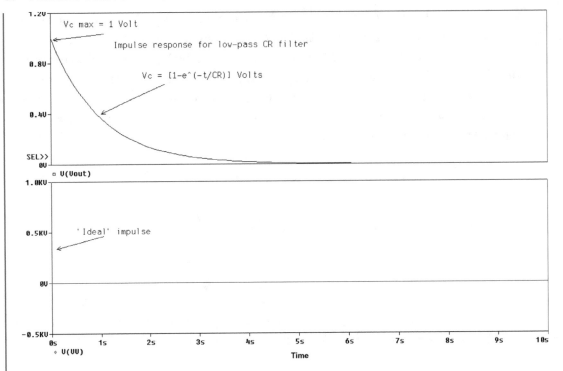

FIGURE 4.11: Impulse response.

4.7.2   Sampling the Impulse Response

Sampling this impulse response by replacing t with nT, or just n, gives the sampled impulse response as

$$h(n) = \omega_c e^{-\omega_c nT}. \tag{4.41}$$

We z-transform (4.41) to yield

$$X(z) = \omega_c \sum_{n=0}^{\infty} e^{-\omega_c nT} z^{-n} = \omega_c \left(e^{-\omega_c 0T} + e^{-\omega_c T} z^{-1} + e^{-\omega_c 2T} z^{-2} + \cdots \right)$$

$$= \omega_c \left(1 + e^{-\omega_c T} z^{-1} + e^{-\omega_c 2T} z^{-2} + \cdots \right). \tag{4.42}$$

We cannot use the equation in this open form, so we need to express this in closed form (i.e., not an infinite series). A general form for (4.42) is $S = a(1 + r + r^2 + r^3 + \cdots)$. Multiply this equation by r to yield $rS = a(r + r^2 + r^3 + \cdots)$ and subtracting the two then gives a more useful form:

$$S - rS = a \Rightarrow S(1 - r) = a \Rightarrow S = \frac{a}{1 - r}. \tag{4.43}$$

To express (4.42) in closed form, we use the result from (4.43), where $a = \omega_c$ and $r = e^{-\omega_c T} z^{-1}$, as

$$H(z) = \frac{\omega_c}{1 - e^{-\omega_c T} z^{-1}}. \qquad (4.44)$$

4.7.3 Mapping from the s-Plane to the z-Plane

This technique maps poles from the s-plane to the z-plane. In this example, an s-plane pole at ω_c maps to a z-plane pole at $z = e^{-\omega_c T}$. With this mapping, the s-plane is mapped in strips $2\pi/T$ wide to the z-plane. A stable analog filter results in a stable digital filter but aliasing may occur because separate poles in the s-plane may map onto the same pole in the z-plane. A good choice for the sampling frequency is 10 times the cut-off frequency. For example, if the analog cut-off frequency is $f_c = 1/(2\pi CR) = 0.159$ Hz, then the sampling frequency is $f_s = 10 f_c = 1.59$ Hz. In Fig. 4.12, the time constant for the analog filter is $\omega_c = 1/\tau = 1 \text{ r s}^{-1} \Rightarrow \tau = CR$ seconds $= 1$ s, so we choose a sampling period $T = 1/f_s = 0.6281$ s, which is the filter delay. Substituting these values into (4.44) yields $e^{-\omega_c T} = e^{-2\pi f_c/f_s T} = e^{-2\pi f_c/10 f_c} = e^{-\pi/5} = 0.5335$, so the digital transfer function is

$$H(z) = \frac{\omega_c}{1 - e^{-\omega_c T} z^{-1}} = \frac{\omega_c}{1 - e^{-2\pi f_c/10 f_c} z^{-1}} = \frac{1}{1 - e^{-\pi/5} z^{-1}} = \frac{1}{1 - 0.5335 z^{-1}}. \qquad (4.45)$$

The passband gain is $20 * log(1/(1 - 0.5335)) = 6.6$ dB. We need to normalize this filter to zero dB by multiplying by $\{1 - 0.5335\}$. **Analysis Set up/Transient** parameters: **Print Step =** 1 s, **Run to time =** 10 s, **Maximum step size =** 0.001. Press **F11** to simulate and produce the analog and digital impulse responses as shown in Fig. 4.13. We see that the shape of the sampled response and the CR response is the same, i.e., the shape is invariant.

The impulse voltage values are measured using the cursor and max icon parameters as follows: (1, 0.5335, 0.2284, 0.151, 0.081, 0.043, 0.025, 0.012). We could press the **FFT** icon

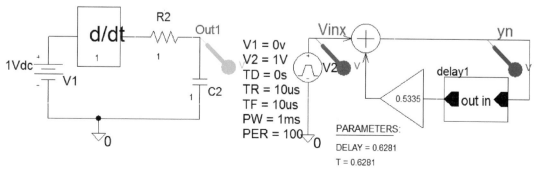

FIGURE 4.12: Digital and analog filters.

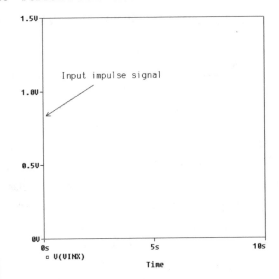
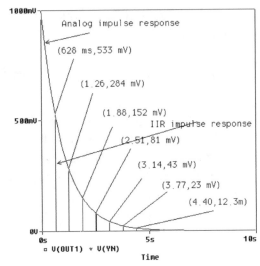

FIGURE 4.13: Sampled impulse response.

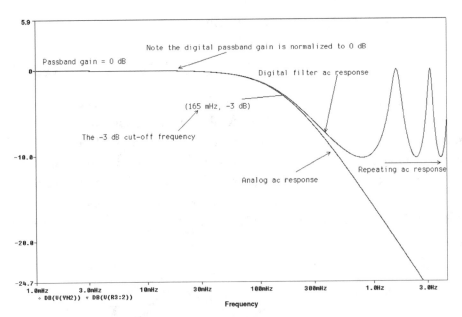

FIGURE 4.14: Frequency response from FFT.

to get the frequency response for the two filters but, instead, we replace the impulse generator with a **VAC** part to produce the AC response shown in Fig. 4.14.

The digital passband gain is 20 log(1/(1−0.5335)) = 6.6 dB, so we need to normalize this to 0 dB to compare the two responses. This requires inserting a **GAIN** part at the input

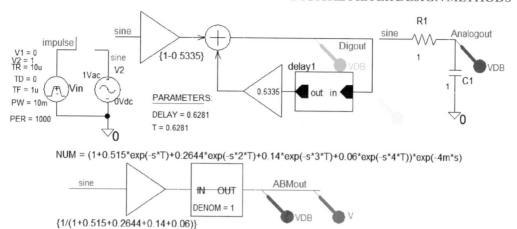

$$NUM = (1+0.515*exp(-s*T)+0.2644*exp(-s*2*T)+0.14*exp(-s*3*T)+0.06*exp(-s*4*T))*exp(-4m*s)$$

FIGURE 4.15: Three filters.

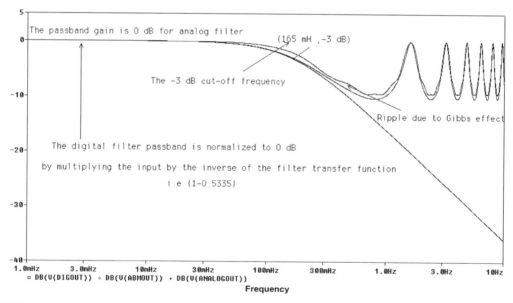

FIGURE 4.16: Frequency response for the three filters.

to the digital filter, with a value $\{1 -0.5335\}$ or just make the **GAIN** part equal to 0.4665. The final part of this analysis is to construct another FIR filter whose coefficients are based on the truncated impulse response. The sampled coefficients are entered into a **Laplace** part with the delays and coefficients as shown in Fig. 4.15. We need to *normalize* the FIR passband gain using a **GAIN** part whose value is set to the inverse of the sum of the filter coefficients, i.e., $\{1/(1 + 0.5335 + 0.284 + 0.151 + 0.081 + 0.043 + 0.023 + 0.0123)\}$.

The frequency response for the three filters is shown in Fig. 4.16. Note the deviation of the digital filter response from the analog response because of aliasing problems.

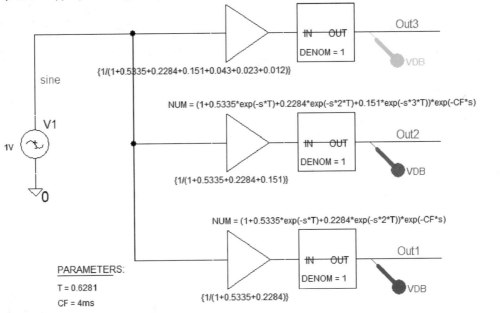

FIGURE 4.17: Change the number of coefficients used.

4.8 TRUNCATING IIR RESPONSES TO SHOW GIBBS EFFECT

The truncated IIR impulse response, with ripple in the transition region, demonstrates Gibbs phenomenon and arises whenever a signal is windowed, or truncated (see the next section). By selecting the first few coefficients in this example, we are effectively truncating the series and so we are using a rectangular window. Fig. 4.17 is a schematic for investigating Gibbs ripple increase when the number of coefficients is reduced (i.e., a reduced window size). CF is a causal factor = 4 ms delay.

We can see from Fig. 4.18 that the passband region is nearly identical, even for a small number of coefficients, but deviates in the transition region. Increasing the number of coefficients improves matters somewhat but at the cost of increased processing and overall delay.

4.9 DESIGNING SECOND-ORDER FILTERS USING THE IMPULSE-INVARIANT METHOD

To apply the impulse-invariant technique to a second-order filter transfer function, we need to use the partial fraction expansion technique. A transfer function is expressed in factored form as

$$H(s) = \sum_{i=1}^{N} \frac{k_i}{s - p_i}. \tag{4.46}$$

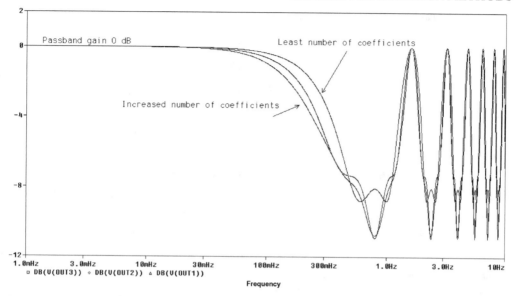

FIGURE 4.18: AC frequency response for different number of taps.

For example, a second-order system has a transfer function defined as

$$H(s) = \frac{1}{(s+1)(s+2)}. \tag{4.47}$$

Apply the partial fraction expansion to (4.47) and break it into two first-order functions:

$$H(s) = \frac{1}{(s+1)(s+2)} = \frac{A(s+2)+B(s+1)}{(s+1)(s+2)} = \frac{s(A+B)+2A+B}{(s+1)(s+2)}. \tag{4.48}$$

Solving (4.48) yields $A = 1$, $B = -1$. Apply the impulse-invariant mapping to each separated function:

$$H(s) = \frac{1}{(s+1)} - \frac{1}{(s+2)}. \tag{4.49}$$

A general form for the impulse response is

$$H(z) = \sum_{i=1}^{N} \frac{k_i}{1 - e^{p_i T} z^{-1}}. \tag{4.50}$$

The digital transfer function is therefore

$$H(z) = \frac{1}{(1 + e^{-T}z^{-1})} - \frac{1}{(1 - e^{-2T}z^{-1})} = \frac{(e^{-T} - e^{-2T})z^{-1}}{1 - (e^{-2T} + e^{-T})z^{-1} + e^{-3T}z^{-2}}. \tag{4.51}$$

For a sampling period $T = \pi/5$, we evaluate (4.51) as

$$H(z) = \frac{0.2489z^{-1}}{1 - 0.8181z^{-1} + 0.151z^{-2}}. \qquad (4.52)$$

Obtain a difference equation from (4.52) and draw a block diagram representation using an IIR filter configuration.

4.10   WINDOWING

Windowing is a technique for truncating a signal to a finite number of sample points. A simple window of length N data points is achieved by multiplying the signal by a rectangular window containing N points. However, Hamming, von Hann (not Hanning!), Chebychev window functions have tapered ends and produce better spectra when compared to spectra using a rectangular window. The window makes the Fourier transform appear as if the data were periodic for all time and is called short-time stationary. Windowing changes the spectral resolution and introduces spectral leakage—an oscillatory signal called Gibbs phenomenon (Josiah Williard Gibbs 1839–1903). When the window contains a nonintegral number of cycles, the end of one windowed signal does not connect to the beginning of the next sample in a continuous manner and so produces glitches at regular intervals. To reduce these glitches, we must window the signal so that the two windowed end amplitudes are nearly zero and connect together more easily.

We will see shortly how windowing is applied to design digital filters. Fig. 4.19 shows the **VPULSE** parameters generating 100 pulses. Set the transient parameters, **Run to time =** 100 initially, and **Maximum step size** = 1 ms, and simulate. To compare the windows, we need to produce a finite block of data. This is achieved using a transmission line to produce a delayed pulse train (the delay is equal to the block length) that is subtracted from the input signal using a **Diff** part. A **PARAM** part defines the window length N and delay k. The block of data is then multiplied by each window function defined in the **ABM1** part.

Select the **ABM** part and **Rclick/Edit Properties** to enter the spreadsheet. A von Hann window function (basically a raised cosine function), is entered as EXP1 = (0.5 + 0.5*cos(2*pi*(TIME-k)/N))*V(%IN). The input block of data V(%IN) is then multiplied by the respective window function. Simulate and observe the different windows and the corresponding spectra. For the Bartlett window we use the (IF, THEN, ELSE) function typed into the **ABM** function to define the function for different time regions, i.e., If(time<=N/2,((2*TIME/N)*V(%IN)),(2-2*TIME/N)*V(%IN)).

We use a delay factor $k = 26$ which is equal to half the window width of $N = 53$. The variable is time, so in the window definition we have (TIME $-k$) instead of n. Use the log function to create a file for separating the window functions into different levels as shown in Fig. 4.20.

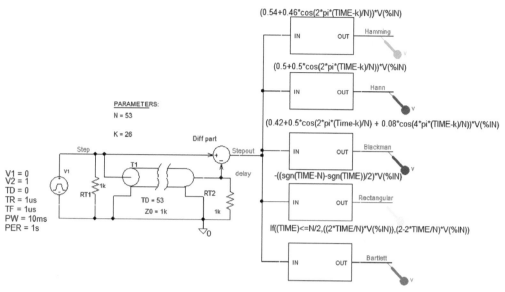

FIGURE 4.19: Windowing.

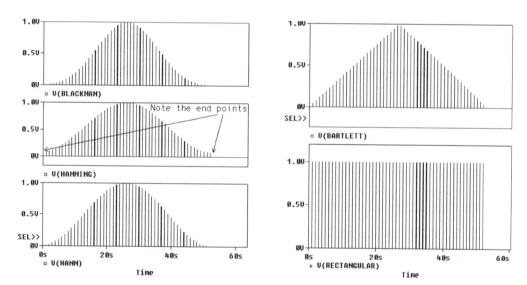

FIGURE 4.20: Window functions.

4.10.1 Windows Plots

Fig. 4.21 shows a magnified section of the spectral plot for each window (use the **FFT** icon + change of axis). The 4 kHz frequency component is represented by a finite-width lobe in both spectral plots, with the rectangular window lobe width equal to $2^* f_s / N = 2 \times 4000/40 = 200$ Hz and the Hamming window lobe length $= 4^* f_s / N = 400$ Hz. The rectangular window

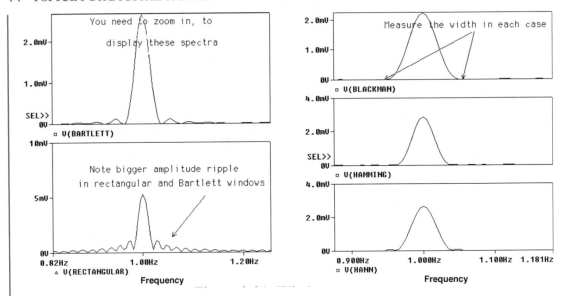

FIGURE 4.21: Window spectra.

gives better resolution in the spectral plots and can resolve closely spaced spectral components when compared to the other windows. Increasing the value of N increases the spectral resolution since the spectral lobe width is inversely proportional to the window length N (see Table 4.1).

The Hamming window is defined as

$$w(n) = \begin{cases} 0.54 + 0.46 \cos(2\pi n/N) & \text{for} \quad -N \leq n \leq N \\ 0 & \text{elsewhere.} \end{cases} \tag{4.53}$$

TABLE 4.1: Window Properties

WINDOW	FUNCTION	TRANSITION WIDTH Δf	RIPPLE (dB)	MAIN LOBE (dB)	STOPBAND (dB)
Rectangular	1	$0.9/N$	0.7416	13	21
Hamming	$0.54 + 0.46 \cos(2\pi n/N)$	$3.3/N$	0.0194	41	53
von Hann	$0.5 + 0.5 \cos(2\pi n/N)$	$3.1/N$	0.0546	31	44
Blackman	$0.42 + 0.5 \cos(2\pi n/(N-1))$ $+ 0.08(4\pi n/(N-1))$	$5.5/N$	0.0017	57	75

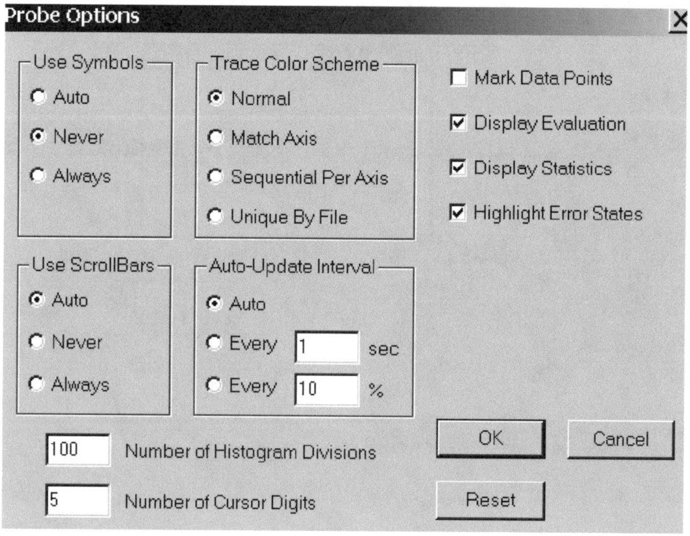

FIGURE 4.22: **PROBE** options.

The size of the window is $2N + 1$. From Fig. 4.21, we can measure the height difference between the major lobe and the next lobe and compare to the values in the above table. However, it will be necessary to increase the **Run to time** to at least 200 ms and simulate. Press the **FFT** icon and apply the magnifying icon to a small section of the spectrum. To display the spectral response in logs, click the **y-axis** icon.

4.10.2 Windows Spectral Plots

We can see how the spectral leakage for a rectangular window is greater than other windows, and hence it tends to mask low-level spectral components. The Hamming window has, however, smaller spectral leakage and is better for measuring signals with a larger spectral range.

Selecting **Tools/Options** produces the display in Fig. 4.22. This menu is used to remove **PROBE** symbols that may clutter the **PROBE** display output after sweeping a parameter.

The **Display Evaluation** is ticked to make a permanent display when using the **Evaluation measurement** functions found in the **PROBE** output menu.

4.11   WINDOW FILTER DESIGN

The window, or Fourier method, is the simplest method for designing FIR filters by truncating an infinite IIR response using a suitable window. The FIR filter coefficients are the magnitudes of the truncated impulse response and achieve a linear phase response filter when the impulse response is *symmetrical*. To realize this filter, we must obtain the impulse response (obtained

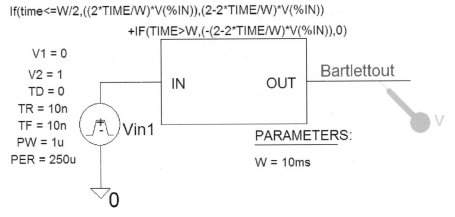

FIGURE 4.23: Producing a Bartlett window.

by getting the inverse Fourier transform of the frequency response), and multiplying it by a window function. The windowing effectively truncates the infinite response but in doing so produces undesirable effects.

4.11.1 Bartlett Window

The Bartlett window is not as important as other windows we will consider, but here we plot the window using the IF, THEN, ELSE statement as If((TIME)<=N/2,((2*TIME/N)*V(%IN)),(2-2*TIME/N)*V(%IN)). Load the schematic shown in Fig. 4.23 and use an **ABM1** part to define the Bartlett window as shown.

The Bartlett window in Fig. 4.24 has $W = 10$ ms with 40 samples.

Matlab has several useful mfiles that you can use to calculate the coefficients. For example, **w = blackman(8)**, which returns the windows coefficients: $w = 0, 0.0905, 0.4592, 0.9204, 0.9204, 0.4592, 0.0905, 0$. Check out the other mfiles Hamming etc.

4.11.2 The Sampled Impulse Response

We investigate a simple digital filter design by making the sample amplitudes from the impulse response of a bandpass filter as the coefficients for an FIR digital filter. The bandpass circuit in Fig. 4.25 is resonant at 73 kHz, and the sampled impulse IIR response is truncated by selecting only a few of the samples. This is the same as applying a rectangular window to produce an FIR-type filter response. The disadvantage of using a rectangular window is that the Gibbs ripple in the frequency response is quite large. A **VPULSE** part, configured as an impulse generator, is applied to the circuit to produce the impulse response. Use the **Net Alias** icon to rename the input wire segment to **Unit impulse**. Set the transient parameters as follows:

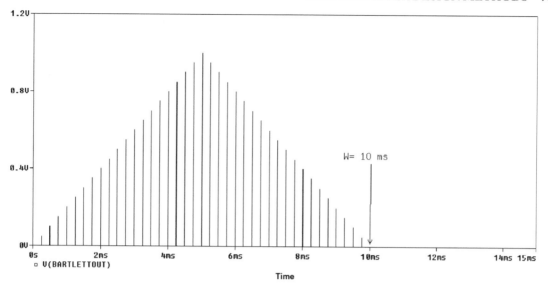

FIGURE 4.24: Bartlett window.

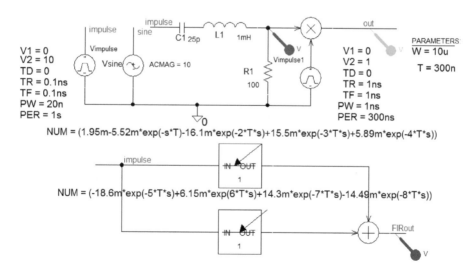

FIGURE 4.25: Analog and digital filters.

Output File Options/Print values in the output file = 20 ms, **Run to time** = 30 ms and **Maximum step size** = 10 ns. Press the **F11** key to simulate.

The idea behind this method is to sample the analog filter impulse response to generate impulses using the sampler with parameters as shown in Fig. 4.26. From the response read the maximum voltage for each impulse and enter them as the digital filter coefficients into the **Laplace** part to define the filter. However, if we enter only a few of these into the **Laplace** part, we effectively are truncating using a rectangular window.

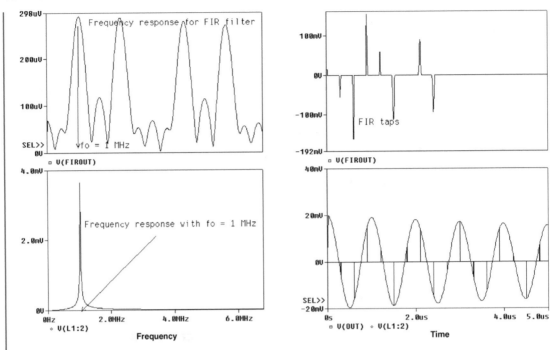

FIGURE 4.26: Impulse response.

The frequency response is obtained by connecting a **VSIN** sine part to the input and renaming the input wire segment to sine. Place a dB marker on the output. From the **Analysis Setup**, select **AC Sweep** and **Linear, Points/Decade** = 1000, **Start Frequency** = 30 kHz, and **End Frequency** = 300 kHz. Press **F11** to simulate. The frequency responses in Fig. 4.27 shows the differences between the analog and digital filters because of the windowed effect of using a few coefficients.

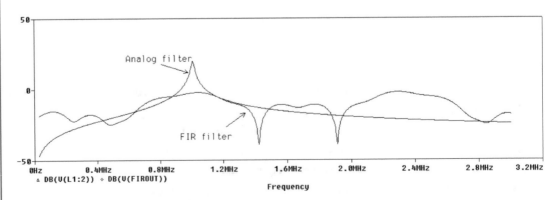

FIGURE 4.27: Impulse invariance filter response.

4.12   IMPULSE RESPONSE OF A BRICK-WALL FILTER

The Fourier transform (FT) is a generalization of the Fourier series used to examine periodic time function, and is used when a signal is not periodic. For example, the FT of a square wave, of duration τ and amplitude A, results in a sinc-shaped spectrum. Consider the following analysis:

$$v(f) = \int_{-\infty}^{\infty} v(t)e^{-j2\pi ft}dt = \int_{-\tau/2}^{\tau/2} Ae^{-j2\pi ft}dt = A\left[\frac{e^{-j2\pi ft}}{-j2\pi f}\right]_{-\tau/2}^{\tau/2}$$

$$= A\tau \frac{(e^{j\pi f\tau} - e^{-j\pi f\tau})}{j2\pi f\tau} = A\tau \frac{\sin \pi f t\tau}{\pi f t\tau} = A\tau \sin c(\pi f\tau). \qquad (4.54)$$

The power spectral density (PSD) is defined as

$$\{v(f)\}^2 = (A\tau)^2 \left(\frac{\sin \pi f\tau}{\pi f\tau}\right)^2. \qquad (4.55)$$

From (4.55), we see that the PSD has a maximum value of $(A\tau)^2$ at 0 Hz (DC), and the first null (a zero crossing) occurs at $\sin \pi f\tau = 0$ (a frequency $f = 1/\tau$), with 90% of the signal energy in the first lobe of the spectrum. As the pulse narrows, the main spectral lobe widens and increases the channel bandwidth requirements in a telecommunications transmiter system. Thus, transmitting infinitely thin digital pulses with no distortion requires a channel with an infinite bandwidth and a linear phase response—conditions that are not physically realizable. Apply the inverse Fourier transform to a narrow pulse processed through an ideal low-pass filter with cut-off frequency $f_c \ll 1/\tau$ to get back to the time domain as

$$v(t) = \int_{-\infty}^{\infty} v(f)e^{j2\pi ft}df = A\tau \int_{-f_c}^{f_c} \frac{\sin \pi ft}{\pi ft}e^{j2\pi ft}df. \qquad (4.56)$$

Use L'Hopital's rule to show that the sinc function $\frac{\sin \pi f\tau}{\pi f\tau}$ has a value of 1 for small values of $f_c\tau$, so (4.56) becomes

$$v(t) = A\tau \int_{-f_c}^{f_c} e^{j2\pi ft}df = A\tau \left[\frac{e^{j2\pi ft}}{j2\pi t}\right]_{-f_c}^{f_c} = A\tau \frac{(e^{j2\pi f_c t} - e^{-j2\pi f_c t})}{j2\pi t}$$

$$= 2Af_c\tau \frac{\sin 2\pi f_c t}{2\pi f_c t} = 2Af_c\tau \sin c2\pi f_c t. \qquad (4.57)$$

A time-domain sinc signal is simulated in Fig. 4.28 using an **ABM** block. Select the **ABM** part, **Rclick/EditProperties** and in the spreadsheet enter the equation 2*A*fc*tau*sin(2*pi*fc*(time-k))/(2*pi*fc*(time-k)) into the **EXP1** box. You could simplify this equation by canceling out the $2fc$ term. The first zero occurs at $\sin 2\pi f_c t = 1$, or $t = 1/2f_c$,

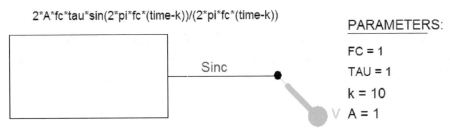

FIGURE 4.28: Sinc pulse.

and other zero crossings at $n/2f_c$. The channel with limited bandwidth causes pulses in the transmitted pulse stream to spread and overlap so that the receiver might not be able to distinguish between 0 and 1. This is called intersymbol interference (ISI) and produces errors. The **PARAM** part is used to define constants including a delay factor, k, included to realize a causal sinc function to make the impulse response noncausal so that it starts from an artificial time $= 0$.

Set the transient parameter **Run to time** $= 100$ s. Press the **FFT** icon and use the magnifying icon to observe the duality that exists between the sinc shaped spectrum and the pulse waveform shown in Fig. 4.29. We can also observe the Gibbs effect occurring in the passband region.

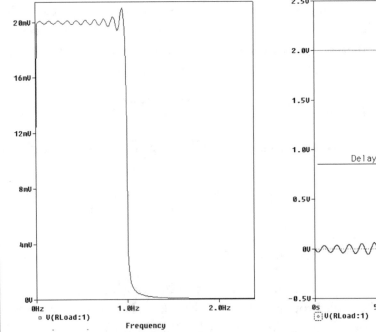

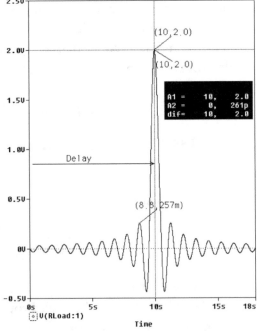

FIGURE 4.29: Sinc signals and pulses.

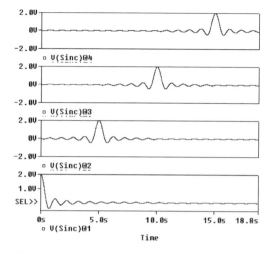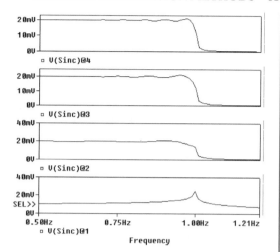

FIGURE 4.30: Varying the delay factor.

Use the Parametric sweep with $k = 0, 5, 15$ to investigate the effect of delaying the sinc signal. There are two main side effects to this technique. The first effect makes the transition region slope become less as the window duration is reduced, as is evidenced by examining the middle right diagram in Fig. 4.30.

The second effect is called the Gibbs phenomenon where ripples occur in the passband whose magnitude is constant and independent of the window. However, ripple frequency is dependent on the window length (compare the two right-hand diagrams). We can use different window shapes for an appropriate transition roll-off rate versus the ripple magnitude.

4.13 DESIGNING FILTERS USING THE WINDOW METHOD

We have seen how truncating an impulse response, and using the samples as FIR filter coefficients, produces filtering. For the filter to operate in real time means placing a limit on the number of coefficients. This windowing produces a response that is not ideal and the limited number of coefficients causes ripples in the passband and a filter with a finite transition region (the width is the same as the major lobe of the window used). Consider the following low-pass filter specification:

- Sampling frequency $f_T = 8$ Hz.
- Passband edge frequency $f_p = 1.5$ Hz.
- Stopband edge frequency $f_s = 2$ Hz.
- Passband ripple $\alpha = 0.1$ dB (or $\delta = 0.0114$) = stopband ripple.

The passband and stopband ripple is the same for this method and the normalized transition frequency is $\Delta f_n = 0.5/8 = 0.0625$ [ref: 5 Appendix A]. The most commonly used window in processing speech is the Hamming window because it avoids the discontinuities associated with a rectangular window (the ends are tapered near zero and make it easier to match the windowed ends to each other). From Table 4.1, we select the Hamming window, which has a transition width $\Delta f = 3.3/N \Rightarrow N = 3.3/\Delta f$, hence $N = 3.3/0.0625 \cong 53$. The number of filter coefficients is therefore 53. If the passband ripple is in dB, then the ripple factor, δ, is

$$\alpha_p = 20\log_{10}(1 - \delta_p) \Rightarrow \delta_p = 1 - 10^{(-\alpha_p/20)} = 1 - 10^{(-0.01/20)} = 0.0114.$$

This is not the ripple specified in the windows table but is the ripple present in the frequency response. The filter order is therefore

$$M = \frac{-20\log(\delta) - 7.95}{14.36(f_s - f_p)/f_T} = \frac{-20\log(\delta) - 7.95}{14.36\Delta f_n} \frac{20\log(0.0114) - 7.95}{14.36(0.5/8)} = 34. \qquad (4.58)$$

Note that a Hann window (Julius von Hann 1839–1921) could have been used in this design. The schematic in Fig. 4.31 uses an **ABM1** part to model a Hamming window and an ideal sinc function. A block of data is created using a transmission delay line part **T**, but an infinite step input is applied to the sinc to produce the infinite sinc response. We make the period of the input signal 1 s, to mimic the index $n = 0, 1, 2, 3$. The filter coefficients are

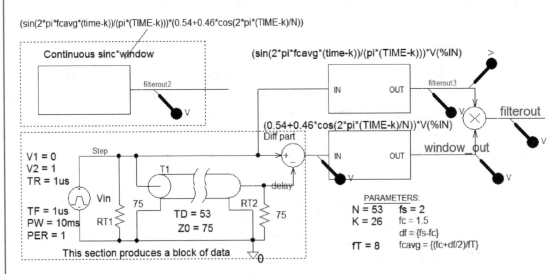

FIGURE 4.31: The window technique for filtering.

calculated by multiplying the window function by the impulse response, i.e.,

$$h(n) = h_d(n).w(n) = 2f_c \frac{\sin(2n\pi f_c)}{2n\pi f_c}\left(0.54 + 0.46\cos\left(\frac{2\pi n}{N}\right)\right)$$

$$= \frac{\sin(2n\pi f_c)}{n\pi}\left(0.54 + 0.46\cos\left(\frac{2n\pi}{N}\right)\right). \tag{4.59}$$

The window smears the cut-off frequency so we calculate a new cut-off frequency as

$$f_c' = f_c + \Delta f/2 = 1.5 + 0.25 = 1.75.$$

Here we use Δf not Δf_n. The normalized average cut-off frequency is $f_{cavg} = 1.75/8 = 0.218$. The coefficients are calculated by substituting values for n. However, we must apply L'Hopital's rule to show that the value for the sinc function at $n = 0$ is 1. Evaluating (4.59) for the first four values of n yields

$$h(0) = (2x0.218)\left(0.54 + 0.46\cos\left(\frac{2\pi 0}{53}\right)\right) = 0.4375$$

$$h(1) = \frac{\sin(2\pi x1 \times 0.218)}{\pi}\left(0.54 + 0.46\cos\left(\frac{2\pi 1}{53}\right)\right) = 0.3109$$

$$h(2) = \frac{\sin(2\pi x2 \times 0.218)}{2\pi}\left(0.54 + 0.46\cos\left(\frac{2\pi 2}{53}\right)\right) = 0.0615$$

$$h(3) = \frac{\sin(2\pi x3 \times 0.218)}{3\pi}\left(0.54 + 0.46\cos\left(\frac{2\pi 3}{53}\right)\right) = -0.0849\ldots \text{ to 53.}$$

$$\tag{4.60}$$

The filter coefficients are symmetrical so we need to compute only half of them. The transient parameters are as follows: **Run to time** = 100 s and **Maximum step size** = 1 ms. Press the **F11** key to produce the response shown in Fig. 4.32.

Click the **FFT** icon to obtain the frequency response from the impulse response. However, to obtain good resolution in the frequency response shown in the right panel in Fig. 4.32, we need to decrease the **Maximum step size** and increase the **Run to time**, which increases the overall simulation time. Remember, what we see from the frequency response is the frequency normalized to the sampling rate of 8 Hz, so that the actual frequency is read as eight times that plotted. It is left as an exercise to construct an FIR filter with the coefficients as calculated and measured, and hence test the filter to see if it meets the specification.

4.14   FIR ROOT-RAISED COSINE FILTER

The transversal FIR filter shown in Fig. 4.33 implements a root-raised cosine filter using a **Laplace** part, where the delay e^{-sT} is entered as exp(-s*T), with T being the sampling period.

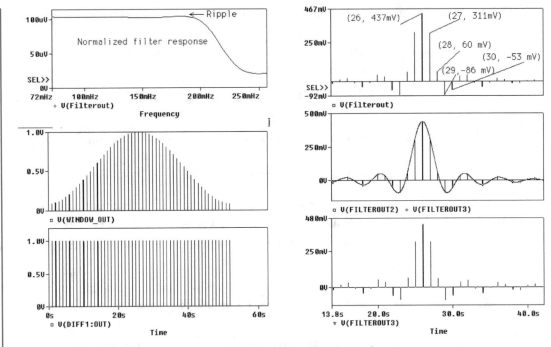

FIGURE 4.32: The frequency response for the windowed and sinc function.

We have to be careful with the delay T value, which is normally set to the symbol duration (the baud time duration). For example, if a symbol length is 1 ms (1000 bps), then the delay T is 1 ms for a sampling rate of 8000 Hz, and not 125 us as it is in a normal digital filter. The 16-tap FIR filter uses **Laplace** part delays, which are not as robust as the transmission delay line but allow us to simulate large filter orders without exceeding the evaluation criteria.

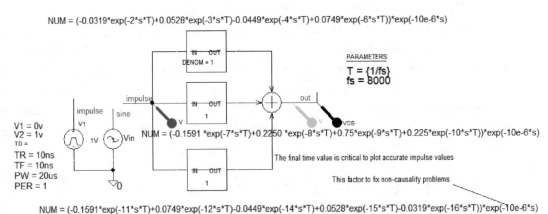

FIGURE 4.33: FIR root-raised cosine filter.

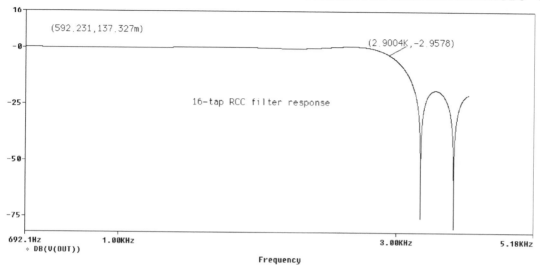

FIGURE 4.34: Raised cosine filter frequency response.

The **Laplace** part is also limited to 132 symbols, so we must use three of these configured as shown.

4.14.1 Raised Cosine FIR Filter Design
A Matlab mfile enables you to design a 16-tap raised cosine filter equalizer by returning the filter coefficients: **b = firrcos(N,F0,DF,Fs)** for an order N low-pass linear phase FIR filter with cut-off frequency F0, sampling frequency Fs, and transition bandwidth DF (all in Hz). For example, **b = firrcos(16,3000,100,8000)** returns $b = -0$, -0.0319, 0.0528, -0.0449 0, 0.0749, -0.1591, 0.225, 0.75, 0.225, -0.1591, 0.0749, 0, -0.0449, 0.0528, -0.0319, -0. Rename the input wire to sine, and from the **Analysis Setup** select **AC Sweep** and **Linear, Points/Decade** = 1000, **Start Frequency** = 100 Hz, and **End Frequency** = 4 kHz. Simulate to produce the response in Fig. 4.34.

Change the input wire segment name to impulse. To plot the coefficients accurately we must enter the following parameters: **Output File Options/Print values in the output file** = 1 us, **Run to time** = 3 ms and **Maximum step size** = 0.1 us. Simulate with the **F11** key to produce the impulse response in Fig. 4.35.

4.14.2 Root-Raised Cosine FIR Filter Design
To design a root-raised cosine filter, use the same mfile but with "sqrt" included: **b = firrcos(16,3000,100,8000, 'sqrt')** returns the coefficients: 0.0034 -0.0342 0.0526 -0.0423 -0.0034 0.0773 -0.159 0.2226 0.7534 0.2226 -0.159 0.0773 -0.0034 -0.0423 0.0526 -0.0342 0.0034. Repeat the simulation for new coefficients.

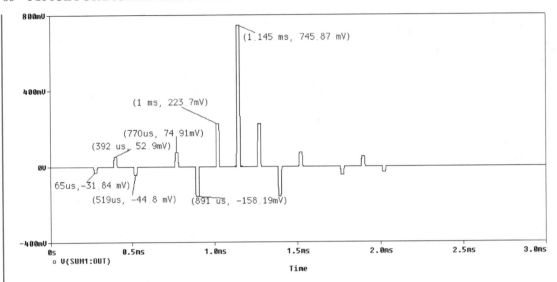

FIGURE 4.35: Impulse response.

4.15 EXERCISES

1. Determine the transfer function for a digital filter that uses the bilinear transform applied to a second-order Butterworth loss function. The transfer function has the form defined as

$$\frac{E_0}{E_i} = \frac{b_p s}{s^2 + b_p s + \omega_p^2}.$$

Here, b_p is the bandwidth, and ω_p is the center frequency (both in r s^{-1}). Show that the resultant digital TF is

$$H(z) = \frac{Y(z)}{X(z)} = \left(\frac{a_0 - a_1 z^{-2}}{1 - b_1 z^{-1} + b_2 z^{-2}} \right).$$

The coefficients are defined as

$$a_0 = a_1 = \frac{2 b_p T}{4 + 2 b_p T + \omega_p^2 T^2},$$

$$b_1 = \frac{2\omega_p^2 T^2 - 8}{4 + 2 b_p T + \omega_p^2 T^2}, \quad \text{and} \quad b_2 = \frac{4 + \omega_p^2 T^2 - 2 b_p T}{4 + 2 b_p T + \omega_p^2 T^2}.$$

2. Investigate the impulse-invariant technique to design a high-pass filter whose cut-off frequency is 1 kHz and which uses a sampling frequency of 10 kHz (i.e., 10 times f_p) and a first-order Butterworth loss function $\$ + 1$. This is not a great method for designing digital high-pass filters because of aliasing problems but worth

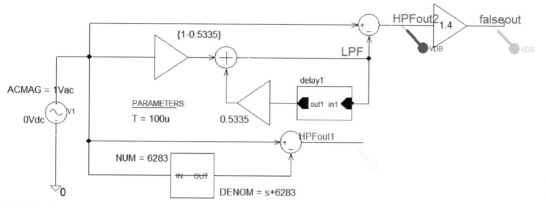

FIGURE 4.36: High-pass filter using the impulse-invariant technique.

investigating, nevertheless. The transfer function for the Laplace part is $H(\$)\big|_{\$=\omega/s} =$ $\frac{1}{\omega/s+1} = \frac{s}{s+\omega} = 1 - \frac{\omega}{s+\omega} = 1 - \frac{6283}{s+6283}$. The transfer function for the equivalent digital filter is $H(z) = 1 - \text{LPF} = 1 - \frac{6283}{1-e^{-\omega_p T}z^{-1}} = 1 - \frac{6283}{1-0.5335z^{-1}}$. We normalize the digital passband gain by multiplying by $(1 - 0.5335)/6283$. The two 8283 factors cancel out and we are left with 0.4665. The 3 dB error in the passband region is "fixed" by inserting a gain of 1.4 as shown in Fig. 4.36.

The right pane in Fig. 4.37 shows the frequency response for the Laplace and digital filters, and a magnified section around the cut-off region is plotted in the left pane.

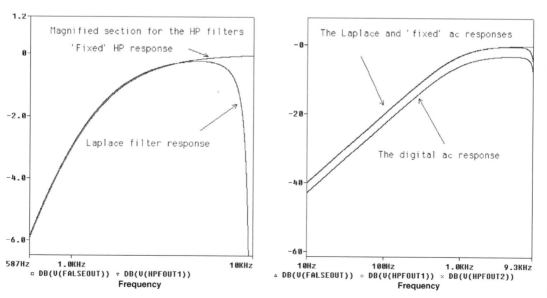

FIGURE 4.37: Amplitude response for Laplace and digital filters.

(1+0.2*sin(2*pi*200*time))*sin(2*pi*4k*(1+0.2*cos(2*pi*300*time))*time)

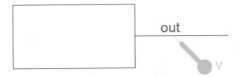

Paplinsky voice model

FIGURE 4.38: Paplinsky voice model.

3. Dual Tone Multi-Frequency (DTMF) signaling comprises the sum of two tones selected from a group of four tones from 630 Hz to 950 Hz and 1200 Hz to 1640 Hz.
Design two IIR filters using the bilinear transform to separate the low tone from the
high tone [ref: 9 Appendix A].

4. Fig. 4.38 shows the Paplinsky model for mimicking the AM and FM characteristics
of speech. Simulate for a few seconds and investigate the spectrum.

CHAPTER 5

Digital Signal Processing Applications

5.1 TELECOMMUNICATION APPLICATIONS

Telecommunication technology has advanced in recent years due to the availability of very powerful digital signal processing integrated circuits. In this chapter, we investigate DSP applications, such as using a FIR filter to implement a Hilbert transformer. An application for the HT is to generate a single-sideband suppressed carrier modulation signal. Music technology devices use DSP devices to produce many musical effects such as reverb, echo, chorus, flanging, etc. to add to, and enhance, the human voice and musical instruments. Such interesting musical effects are easily achieved using variations of FIR and IIR filters with feedback and feed-forward delays and variable delays, etc. The first DSP application is the production of sinusoidal signals that have a 90° phase difference between them.

5.2 QUADRATURE CARRIER SIGNALS

Passband modulating techniques use cosine and sine quadrature carriers to produce multilevel (M-ary) passband signals [ref: 2 Appendix A]. Quadrature signals $y(n) = \cos n\theta$, and $x(n) = \sin n\theta$ are generated using the system shown in Fig. 5.1. Since the output is $y(n) = \cos(n\theta)$, then $y(n+1) = \cos(n+1)\theta = \cos(n\theta + \theta) = \cos n\theta \cos \theta - \sin n\theta \sin \theta$. Substituting $n-1$ for n yields the cosine output as

$$y(n) = \cos(n\theta) = \cos(n-1)\theta \cos \theta - \sin(n-1)\theta \sin \theta. \tag{5.1}$$

Similarly, for the sine output

$$x(n) = \sin n\theta = \sin(n-1)\theta \cos \theta + \cos(n-1)\theta \sin \theta. \tag{5.2}$$

We can draw the block diagram using these equations by starting at the output and working backward. The two delays are identical and are implemented using a transmission line part.

Run to time = 5 ms, and **Output File Options/Print values in the output file** = 1 us. Press the **F11** key to simulate. We need to "kick-start" the oscillator using an impulse signal to produce the display in Fig. 5.2.

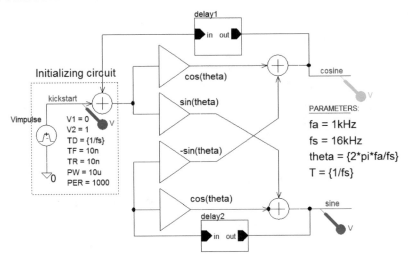

FIGURE 5.1: Quadrature oscillator.

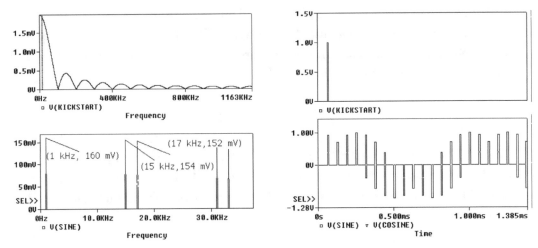

FIGURE 5.2: Quadrature output signals.

The cosine signal was copied to a separate axis using **alt PP, ctrl C**, and **ctrl V**. Press the **FFT** icon to display the spectrum for the cosine output. Since this is a sampled cosine, then the spectrum contains sidebands centered around the 16 kHz sampling frequency, and also at multiples of 16 kHz.

5.3   HILBERT TRANSFORM

The Hilbert transformer (HT) has many DSP applications and our first application uses it to introduce a constant 90° phase shift to a complex signal whose spectrum contains a band of

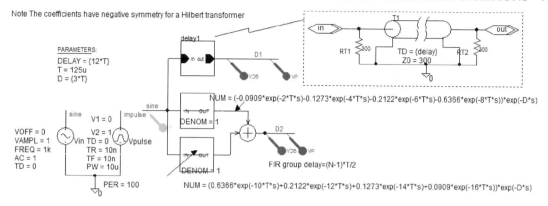

FIGURE 5.3: Hilbert transformer.

frequencies [ref: 3 Appendix A]. Fig. 5.3 shows a multitap FIR implementing an HT whose filter output is in quadrature with a delay expressed as

$$\text{Delay} = (N-1)T_s/2 = 16 \times 125u/2 = 1000 \text{ us.} \qquad (5.3)$$

The FIR filter has an odd number of coefficients (or taps) $N = 17$ but half of them are zero (it cuts down on the computational load). A delay of D = 3*T is added to the delay and filter to fix causal problems. A little bit of tweaking to the delay was necessary in order to produce the required 90° phase shift (use the **PARAM** part with the delay variable from 1.1 ms to 1.2 ms plus D). The **delay1** of 1.5 ms is defined in a **PARAM** part and a differential voltage marker pair and voltage phase markers are attached as shown. The **ABM** blocks have a limited number of characters so we need to use two of them connected as shown. The nonzero filter coefficients and delays are defined in two **ABM** blocks as follows:

Block 1

(−0.0909*exp(−2*T*s)−0.1273*exp(−4*T*s)−0.2122*exp(−6*T*s)
−0.6366*exp(−8*T*s))*exp(−D*s)

Block 2

(0.6366*exp(−10*T*s)+0.2122*exp(−12*T*s)+0.1273*exp(−14*T*s)
+0.0909*exp(−16*T*s))*exp(−D*s)

The coefficients in each block are multiplied by exp(-D*s) to overcome causal problems.

The odd-numbered coefficients are zero, which reduces the number of multiplications required and thus cuts down on the number of computations. An impulse is applied to the HB by changing the input wire segment name to **impulse**. The **Transient Analysis** parameters are as follows: **Output File Options/Print values in the output file** = 100 ns, **Run to time** = 4 ms and **Maximum step size** = 0.1 us. One technique for determining the filter coefficients is to use a Matlab mfile, e.g., **b = remez(16,[0.1 0.9],[1 1], 'hilbert')** which returns: −0.0001, −0.053, 0, −0.0882, 0, −0.1868, 0, −0.6279, 0, 0.6279, 0, 0.1868, 0, 0.0882, 0, 0.053, 0.0001. Note that the odd coefficients are zero.

5.3.1   The Hilbert Impulse Response

After pressing **F11**, you should observe in Fig. 5.4 how each delayed impulse has the same magnitude as the corresponding filter coefficient. Observe the symmetry in the display, and note how the delayed impulse is located at the center of the FIR impulse response.

The complete delay is shown in Fig. 5.5. Place phase and dB markers on the wire segments **D1** and **D2** and change the input wire name to **sine**. Carry out an AC analysis by setting the **Analysis Setup** parameters as follows: **AC Sweep/Linear, Points/Decade** = 10001, **Start Frequency** = 10 Hz and **End Frequency** = 8 kHz.

5.3.2   The Hilbert Amplitude and Phase Responses

Press **F11** to simulate and produce the frequency and phase responses shown in Fig. 5.5. Note that we plot the phase variable as **VP(D2)-VP(D1)**, and the AC response as **dB(V(D1,D2))**.

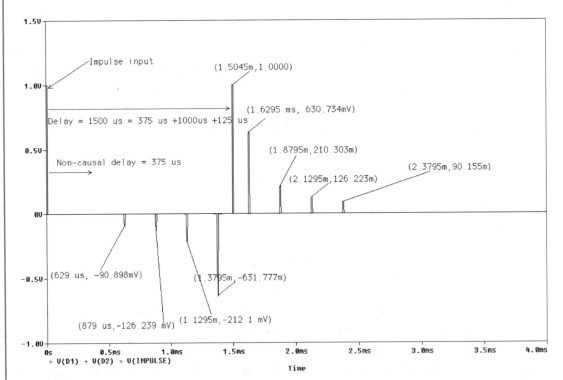

FIGURE 5.4: The impulse response.

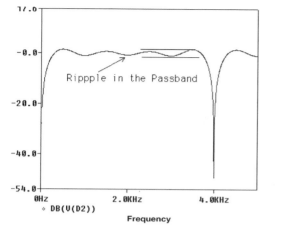 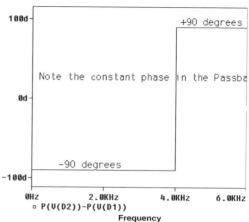

FIGURE 5.5: Frequency and phase response of the Hilbert transform.

Even though you have selected **Linear** in the **Analysis Setup**, you might have to select the **x-axis** and reset it to **Linear**. A bandpass response occurs between 0 and $f_s/4$, and the phase is $-90°$ over this range. However, a $+90°$ is achieved by changing the order of the phase markers, i.e., **VP(D1)-VPD(2)**. From the **PROBE/File** menu tick **Run Commands** and select **Fig. 5-007.cmd** to automatically produce the keystrokes necessary for the correct display. Ripple is present in the passband but this is reduced by increasing the number of taps (the filter order).

5.4 SINGLE-SIDEBAND SUPPRESSED CARRIER MODULATION

In a double-sideband suppressed carrier (DSBSC) system [ref: 2 Appendix A], the carrier is suppressed and increases the transmission efficiency and the signal to noise ratio, when compared to double-sideband full carrier (the same transmitted power in each). Both sidebands contain the same information, so that eliminating one of the sidebands gives a further increase in the overall efficiency. This is called single-sideband suppressed carrier (SSBSC) and can be implemented in several ways. The filter method is the simplest technique to understand (one of the sidebands is eliminated using a very high Q-factor ceramic filter) but another technique, called the phase-shift method (see Exercise 6), is the sum of two products, the baseband signal and the carrier, and the same two signals but shifted by $90°$ using the Hilbert transform. In Fig. 5.6, two sinusoidal modulating signals are applied to test the SSB modulator (a vector modulator). The output signals from the two multipliers are then subtracted in a **Diff** part.

Change the input wire segment to **twosines** and set the **Transient Analysis** parameters as follows: **Run to time** = 4 ms and **Maximum step size** = 0.1 us, and simulate. From the **PROBE/Run Commands** menu select **Fig.5-007.cmd** to automatically separate and display the variables. Fig. 5.7 shows how the carrier and upper sidebands are now eliminated in the final output.

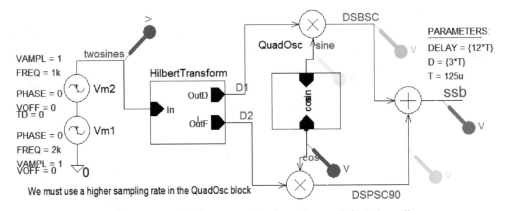

FIGURE 5.6: SSB modulator using Hilbert transformer and a quad digital oscillator.

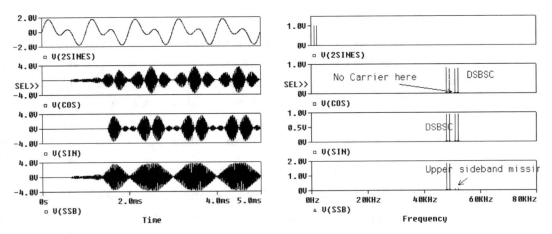

FIGURE 5.7: Testing SSB using two sinusoids.

5.5 DIFFERENTIATOR

A single time delay is represented in the z-domain by multiplying the transformed signal by z^{-1}, i.e.,

$$Y(z) = X(z) - X(z)z^{-1} = X(z)(1 - z^{-1}) = X(z)(1 - e^{-s\tau}). \tag{5.4}$$

If the delay is much shorter than the input pulse width, i.e., $\omega\tau \ll 1$ or $\omega \ll 1/\tau$, then the triangular waveform at the input to the differentiator will appear as a square wave at the output because the output is approximately

$$Y(z) \approx X(z)[1 - (1 - s\tau)] \approx X(z)(s\tau) = X(z)(j\omega\tau). \tag{5.5}$$

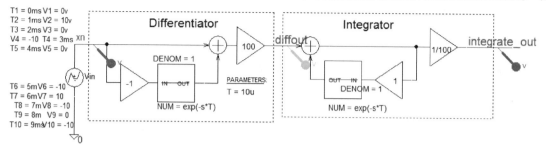

FIGURE 5.8: Differentiator and integrator.

5.6 INTEGRATOR

The final output is

$$Y'(z) = Y(z) + Y'(z)z^{-1} \Rightarrow Y(z) = Y'(z)(1 - z^{-1}) \Rightarrow Y'(z) = Y(z)/(1 - e^{-st}). \quad (5.6)$$

If the delay is much shorter than the input pulse width, i.e., $\omega\tau \ll 1 \Rightarrow \omega \ll 1/\tau$, we see that the final output is

$$Y(z) \approx Y(z)/[1 - (1 - s\tau)] \approx Y(z)/(s\tau). \quad (5.7)$$

The differentiator and integrator circuits are shown in Fig. 5.8.

Set the **Transient Analysis** parameters as follows: **Run to time** $= 50$ ms, **Maximum step size** $= 1$ us. Press **F11** to simulate and plot the integrated, differentiated, and recovered input signals displayed in Fig. 5.9. From the **PROBE File** menu, tick **Run Commands** and select **Figure5-008.cmd** to automatically produce the keystrokes for the correct display. We see how the low-pass filtering removes the high frequencies from the input signal.

5.7 MULTIRATE SYSTEMS: DECIMATION AND INTERPOLATION

Antialiasing filters are inserted to attenuate components in the input signal spectrum whose frequencies are above half the sampling rate, f_s. If the filter has a cut-off frequency of $f_s/2$, then the system satisfies the Shannon–Nyquist criterion, thus eliminating aliasing frequencies. A higher than theoretically necessary sampling rate means that we can use a lower order antialiasing filter but this higher sampling rate introduces extra processing in the digital signal processor. To reduce this processing load and increase efficiency, we introduce multirate sampling (different sampling frequencies at different parts of the system). Another reason for multirate sampling is to match sampling frequencies when transferring data between systems with different sampling rates. For example, transferring data between a domestic CD audio system with a sampling

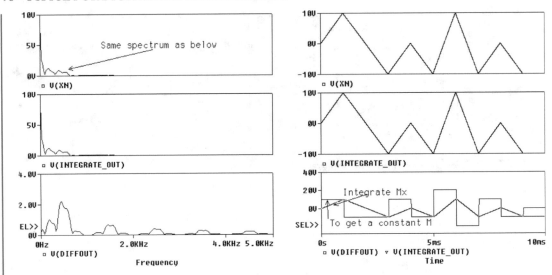

FIGURE 5.9: Differentiator and integrator signals.

rate of 44.1 kHz, and a professional audio system with a sampling rate of 48 kHz, requires changing the sampling frequency rate. This produces a noninteger sampling rate calculated as $44100/48000 = 147/160 = 0.91875$ and requires that we interpolate by a factor of 147 followed by a decimating factor of 160. Multirate sampling consists of two main processes:

- *Decimation*: to decrease the sampling rate.
- *Interpolation*: to increase the sampling rate.

5.8   DECIMATION

Decimation is a filter and down-sample technique for lowering the sampling rate by reducing the number of samples. For example, decimating by 2 eliminates every other sample thus halving the sampling frequency. However, to satisfy the Shannon–Nyquist rate and prevent aliasing components corrupting the baseband signal, we need to put the signal through a FIR digital low-pass filter before decimating takes place. The deciMation factor, M, is the ratio of the input rate to the output rate and reduces the signal processing by $1/M^2$. The output signal after decimating the input signal $x(n)$ is $y(n/M)$. The decimator circuit in Fig. 5.12 shows a single sinusoidal signal sampled at 8 kHz and is contained in the **Signals** block.

Decimation is achieved by sampling the sampled signal at a lower sampling frequency. An FIR digital antialiasing filter is used in decimating circuits. We can design an elliptical filter using the Matlab mfile **[b,a] = ellip(N,Rp,Rs,Wn)**, where N is the filter

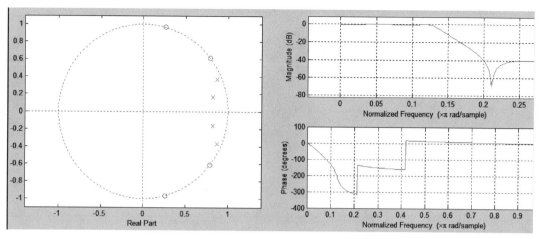

FIGURE 5.10:

order, R_p is the passband in dB, R_s is the stopband in dB, and W is the cut-off frequency/2 times sampling frequency. **[b,a] = ellip(4,1,50,1000/(2*fs))** returns the numerator coefficients $b = 0.0055 - 0.0087\ 0.0121 - 0.0087\ 0.0055$, and the denominator coefficients $a = 1.0000 - 3.4799\ 4.6596 - 2.8977\ 0.6904$. The pole-zero and frequency responses were plotted in Fig. 5.10 using zplane(b,a) and freqz(b,a).

Connect the vac sine generator to the filter input and simulate to produce the response shown in Fig. 5.11.

The insides of the three blocks are shown in Fig. 5.13.

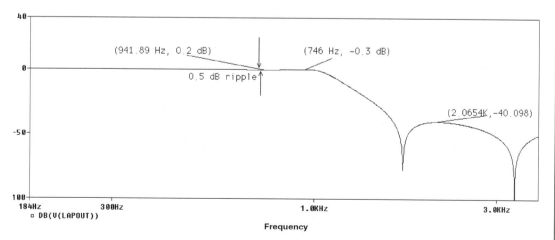

FIGURE 5.11:

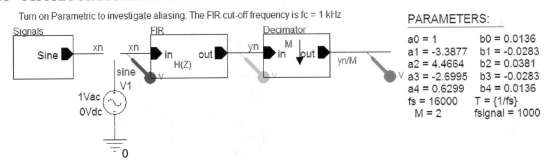

FIGURE 5.12: Decimating by a factor M.

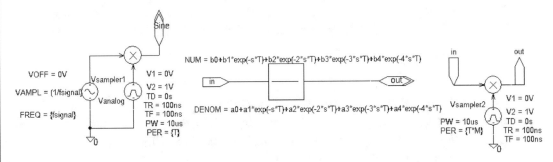

FIGURE 5.13: The decimator circuits.

To create an input signal with a range of frequencies and amplitudes, we make the **VSIN** part parameters **VAMPL** $= \{1/\text{fsignal}\}$, **FREQ** $= \{\text{signal}\}$. From the **Analysis Setup** menu select **Parametric** and tick **Global Parameter and Linear** and set **Name** $=$ fsignal, **Start Value** $=$ 1k, **End Value** $=$ 5k, and **Increment** $=$ 2k. The decimation factor M is defined in the **PARAM** part, and the decimating sampler period is set at M times the original sampling rate as **PER** $= \{\text{T*M}\}$. The filter stopband edge frequency is also reduced by M to attenuate aliasing components. The new stopband edge frequency is calculated as

$$f_s = F_{\text{new\_output\_sample\_rate}} - \frac{F_{\text{original\_sample\_rate}}}{2M}. \tag{5.8}$$

For example, if the original sampling frequency is 16 kHz and the new sampling frequency is 8 kHz then $M = 2$, hence the new stopband edge frequency is

$$f_{si} = f_i - \frac{f_s}{2M} = 8 \text{ kHz} - \frac{16}{2 \times 2} = 4 \text{ kHz}. \tag{5.9}$$

It is better to decimate in stages where large decimation factors are required, because the processing and storage efficiencies are improved. We may compare the efficiency of multirate systems by considering the total number of multiplication per second (MPS) [ref: 6 Appendix A],

$$\text{MPS} = \sum_{i=1}^{L} N_i F_i,$$ (5.10)

where i is a dummy variable representing a particular stage, N_i is the stage coefficient number, and F_i is the stage sampling rate. The total storage requirement is calculated as

$$\text{TSR} = \sum_{i=1}^{L} N_i.$$ (5.11)

We may use these two equations to compare the efficiency of decimating by a large factor, or in stages. For example, if we require a decimating factor of 64, then we could use three stages 8, 4, 2 (i.e., $64 = 8 \times 4 \times 2$). Observe the decimated output signal in Fig. 5.14. From the **PROBE File** menu tick **Run Commands**, and select **Figure5-013.cmd** to automatically produce the keystrokes necessary for the correct display.

Click the **FFT** icon and apply the magnifying icon to examine the initial part of the sinc signal in Fig. 5.15. Here we see how the new decimated signal spectra show sidebands now located at 8 kHz instead of 16 kHz.

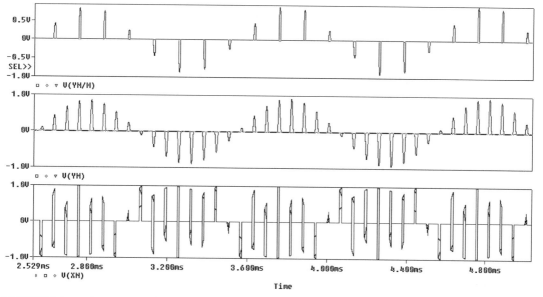

FIGURE 5.14: Decimation signals.

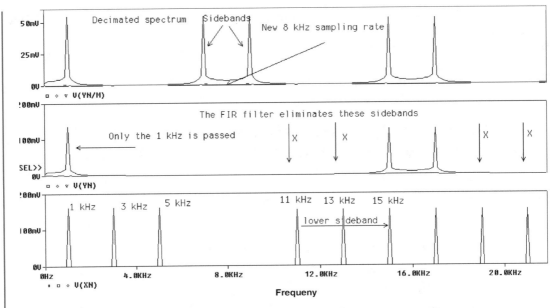

FIGURE 5.15: Spectrum of decimation signals.

5.8.1 Example

a) Design a two-stage decimator to down-sample a sampled audio signal originally filtered to 3.4 kHz. The reduction is from 240 kHz to 8 kHz with one of the decimating factors being 15. The decimator specification is as follows:

- Sampling frequency $f_T = 240$ kHz.

- Passband edge frequency $f_p = 3.4$ kHz.

- Peak passband ripple $\delta_p = 0.025$.

- Minimum stopband ripple $\delta_s = 0.01$.

The stopband edge frequency is

$$f_s = F_{\text{new\_output\_sample\_rate}} - \frac{F_{\text{original\_sample\_rate}}}{2M}.$$

The filter order is calculated as

$$N = 1 - \frac{10\log(\delta_p\delta_s) + 13}{14.6\frac{f_s - f_p}{F_T}}.$$

b) Determine the number of multiplications per second $\text{MPS} = \sum_{i=1}^{L} N_i F_i$ and the total storage requirements.

5.8.2   Solution

The decimator factor $M = 240$ kHz/8 kHz $= 30$, so the decimating factors for the two stages are 15 and 2. The output sample rate for the first stage is $240/15 = 16$ kHz and the second output sampling rate is 8 kHz. The first-stage filter has the same passband edge frequency $= 3.4$ kHz, but the stopband edge frequency is

$$f_s = 16 \text{ kHz} - \frac{240 \text{ kHz}}{2 \times 30} = 12 \text{ kHz}.$$

The normalized transition width is

$$\Delta f = \frac{f_s - f_p}{f_T} = \frac{12 \text{ kHz} - 3.4 \text{ kHz}}{240 \text{ kHz}} = 0.0358.$$

The filter order is

$$n = 1 - \frac{10 \log(\delta_p \delta_s) - 13}{14.6 \Delta f} = 1 - \frac{(10 \log(0.025 \times 0.01) + 13)}{14.6 \times 0.0385} = 45.$$

The second-stage stopband edge frequency is

$$f_s = 8 \text{ kHz} - \frac{240 \text{ kHz}}{2 \times 30} = 4 \text{ kHz}.$$

The normalized transition width is

$$\Delta f = \frac{f_s - f_p}{F_T} = \frac{4 \text{ kHz} - 3.4 \text{ kHz}}{16 \text{ kHz}} = 0.0375.$$

The filter order for the second stage is

$$n = 1 - \frac{10 \log(\delta_p \delta_s) + 13}{14.6 \Delta f} = 1 - \frac{(10 \log(0.025 \times 0.01) + 13)}{14.6 \times 0.0375} = 43.$$

The total number of multiplications per second MPS $= \sum_{i=1}^{L} N_i F_i = 45 \times 16 + 43 \times 8 = 1064$. The total storage requirement is calculated as TSR $= \sum_{i=1}^{L} N_i = 45 + 43 = 88.$

5.9   ALIASING

To demonstrate the aliasing phenomenon we must redesign the filter bandwidth so that it passes the higher 5 kHz frequency such that the Nyquist rate is not satisfied i.e. $f_s < 2 f_m$ at the new sampling frequency of 8 kHz. Redesign the filter with a new cut-off frequency $f_p = 5$ kHz using $[\mathbf{b}, \mathbf{a}] = \mathbf{ellip(4, 0.5, 40, 5000/(0.5*16000))}$ to return $\boldsymbol{b} = 0.2094, 0.7049, 1.0036, 0.7049, 0.2094$, and $\boldsymbol{a} = 1.000, 0.7296, 0.9783, 0.1412, 0.1508$ as the new coefficients and simulate. The new filter frequency response is shown in Fig. 5.16.

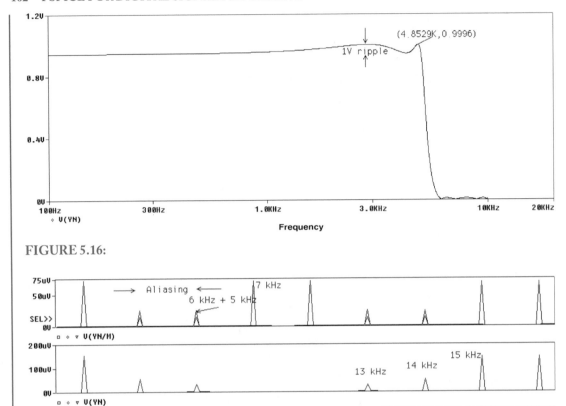

FIGURE 5.16:

FIGURE 5.17: Observe the overlap/aliasing distortion.

5.10 INTERPOLATION

Interpolation is a technique to increase the sampling rate to match systems operating at a higher sampling rate but the spectrum remains unchanged. Interpolation allows a simpler reconstruction filter design at the output of a digital to analog converter (DAC). Up-sampling, by inserting zero-valued samples between original samples (called zero stuffing) increases the sampling rate but adds undesirable spectral images to the original signal that are at multiples of the original sampling rate. Aliasing is different to spectral imaging because aliasing causes a loss of information. However, up-sampling must be followed by filtering to remove these spectral images. Note that in this simulation, the samples added should be zero but are given small amplitudes in order to see them. To show how the spectrum is unchanged after interpolation, consider the z-transform of the sequence $x(n) = \{1, 2, 3, 4\}$ before interpolation,

$$X(z) = \sum_{n=0}^{3} x(n)z^{-n} = z^{-0} + 2z^{-1} + 3z^{-2} + 4z^{-2} = 1 + 2z^{-1} + 3z^{-2} + 4z^{-3}, \qquad (5.12)$$

where $z = e^{sT} = e^{j\omega T}$, and T is the sampling period equal to the inverse of the sampling frequency. If we interpolate the sequence with $L = 1$, then we insert a zero between existing samples so we rewrite (5.13) with $z_L = e^{sT/2} = e^{j\omega T/2}$:

$$X_L(z_L) = \sum_{n=0}^{3} x(n)z_L^{-n} = 1 + 0z_L^{-1} + 2z_L^{-2} + 0z_L^{-3} + 3z_L^{-4} + 0z_L^{-5} + 4z_L^{-6}$$
$$= 1 + 2z^{-1} + 3z^{-2} + 4z^{-3}. \qquad (5.13)$$

5.11   DECIMATION AND INTERPOLATION FOR NONINTEGER SAMPLING FREQUENCIES

Load the schematic in Fig. 5.18. Interpolating and filtering produces the same result as if you had originally sampled your signal at the higher rate. The interpoLation factor, L, is the ratio of the output sampling rate to the input sampling rate. To achieve noninteger interpolation factors, you must combine interpolation and decimation, where $M = 2$ and $L = 3$. The image removal filter after the decimator eliminates the images created by interpolation, and has a stopband edge frequency $Lf_s/2$. Thus, if the input sampling rate is 44 kHz and you wish to interpolate by 2, i.e., $L = 2$, then the stopband edge frequency is 2×44 kHz/$2 = 44$ kHz.

The extra padded 'zeroes' amplitudes are set to a finite value such as 0.01 volts. The zeroes should be zero but we make them finite in order to see them in the interpolated signal. Set the transient analysis parameters: **Run to time** = 20ms and **Maximum step size** = 100ns and simulate. From the **Probe/Run Commands menu** select **Fig.5-020.cmd** to automatically separate and display the variables into a less confusing display. Figure 5.20 shows the decimated signal in the middle right pane and the interpolated signal with the extra added padding samples. The decimator sampling period **PER** is set to $\{(1/fs)*M\}$, and the interpolator sampling rate

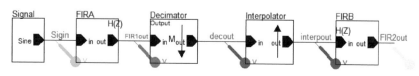

FIGURE 5.18: Decimation and interpolation.

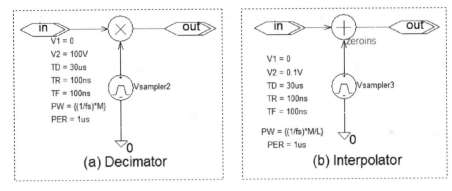

FIGURE 5.19: The decimator.

is $\{(1/fs)*M/L\}$ with the padded zeros having an amplitude of 0.1 V in order to see the extra components (they should be zero).

The decimated signal in the middle one and the interpolated signal where the extra padding sampling signals are added albeit reduced in amplitude but not zero.

The signals for the decimator and interpolator in the frequency domain in the left pane of Fig. 5.20.

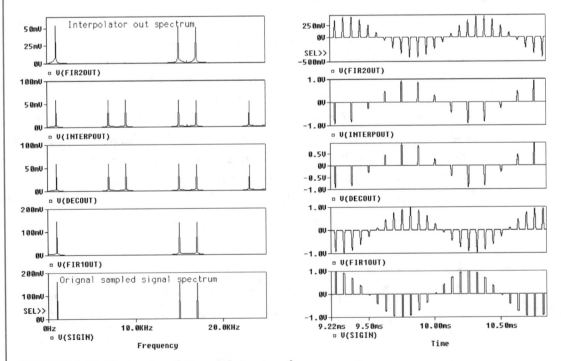

FIGURE 5.20: Decimation and interpolation waveforms.

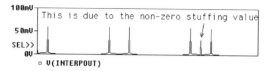

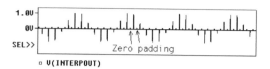

FIGURE 5.21: Increasing the zero stuffing to 0.1 V.

5.12   EXERCISES

1. Investigate the effect of the coefficients on the integrator filter (LPF) shown in Fig. 5.22.

2. The integrator action is shown in Fig. 5.23. Connect the **VSIN** generator and obtain the frequency response.

3. Investigate the digital filter in Fig. 5.24, where two sampled signals are applied to the input.

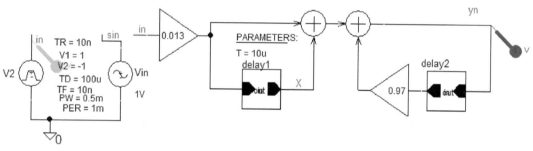

FIGURE 5.22: Digital integrator.

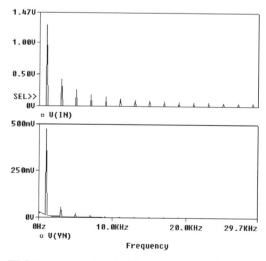

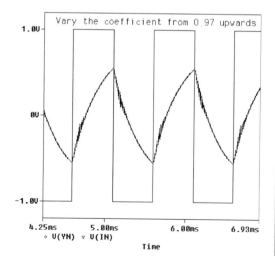

FIGURE 5.23: Digital integrator waveforms.

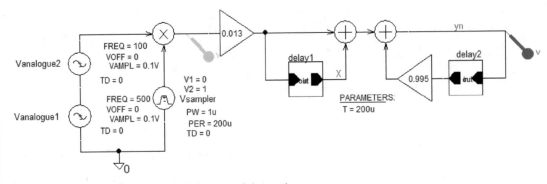

FIGURE 5.24: Applying a sampled sinusoidal signal.

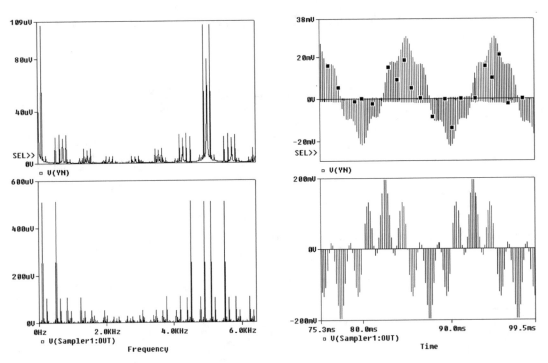

FIGURE 5.25: Recovered signal and spectrum.

The 500 Hz sideband is attenuated as shown in the zoomed-in top left pane in Fig. 5.25.

4. Investigate the all-pass filter in Fig. 5.26. The filter difference equation is

$$Y(z) = 0.7X(z) - 0.99X(z)z^{-1} + 0.7Y(z)z^{-1}.$$

Hence, the transfer function is

$$H(z) = Y(z)/X(z) = (0.7 - 0.99z^{-1})/(1 - 0.7z^{-1}).$$

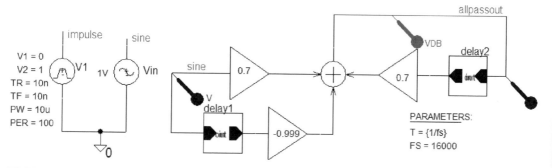

FIGURE 5.26: All-pass filter.

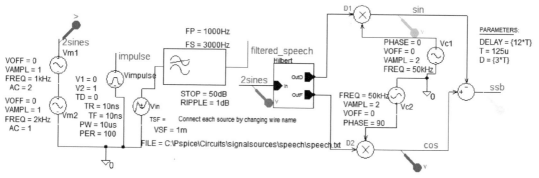

FIGURE 5.27: SSBSC generation using digital quad oscillators.

Make sure to change the **x-axis** to linear and set **Maximum step size** to 0.1 us. Note: change the output **y-axis** 0 dB to −50 dB.

5. Fig. 5.27 shows an SSBSC modulator using quadrature oscillators. The quadrature sine–cosine oscillators examined previously are located in the **QuadOsc** block and produce quadrature DSBSC outputs. Apply a speech signal to the SSB modulator and investigate the final output SSBSC spectrum.

CHAPTER 6

Down-Sampling and Digital Receivers

6.1   RECEIVER DESIGN

Conventional AM/FM receivers have been around for a long time and, apart from the introduction of semiconductor technology, have not changed much from Armstrong's original design. Digital receivers are divided into wideband and narrowband depending on the decimation range and use one of the following methods:

- RF sampling method.

- Direct Conversion method.

- IF undersampling method.

Existing digital receivers are a mixture of analog and digital techniques, because the RF signal from the aerial is still in analog form. The RF signal is frequency-translated downward and then digitized at RF, or IF, where DSP techniques are applied to recover the baseband signal. The direction conversion method was popular because existing analog to digital converters were unable to operate at high frequencies and could only digitize the lower-frequency baseband signals. The IF undersampling method samples the IF output and uses a sampling rate that is at least twice the highest frequency in the IF signal to prevent aliasing effects. Intentionally undersampling the signal at a frequency lower than the Nyquist rate aliases the signal to the desired baseband frequency range [ref: 9 Appendix A].

6.2   RF SAMPLING

Fig. 6.1 shows the essential elements of a digital receiver using the RF sampling method. Here, the analog to digital converter (ADC) samples the initial bandpass filter output with the Nyquist sampling frequency determined by the maximum input frequency. For example, if the maximum input frequency is 10 MHz, then the sampling rate must be at least 20 MHz. The baseband signal does not require a high sampling rate, so the signal is decimated and then the signal is detected to recover the original modulating signal. The complex mixer consists

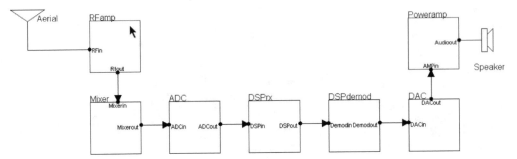

FIGURE 6.1: Digital receiver.

of two digital multipliers that accept digital samples from the ADC, and the local oscillator. They produce a complex representation of the input signal, which has been translated down by the local oscillator. By appropriately tuning the local oscillator, any frequency band of interest can be centered on 0 Hz. The decimator FIR low-pass filter accepts I and Q samples from the mixer. At the filter output, a decimation stage drops all but one of every N samples.

We will investigate parts of this receiver and, as before, we use sampled signals but not in digital form, i.e., not 1s and 0s from the ADC. In a real digital receiver, the output from the ADC is in parallel form where each bit is then multiplied digitally (shift and add). Dedicated DSP-integrated circuits are similar to standard microprocessors and perform a multiplication as a single instruction in a multiply and accumulate (MAC) device. For example, two pieces of data are multiplied together and added to a third piece of data in an accumulator.

6.2.1   Down-Sampling a Passband Signal

In Fig. 6.2 we use an **ABM** part to define a single sideband suppressed carrier modulated by two modulating frequencies at 1 kHz and 5 kHz. The upper side frequencies are located to the right of the 1 MHz carrier. This SSBSC is sampled at 20 kHz and results in the two modulating signals appearing at baseband. Down-converting thus replaces the local oscillator and mixer. At first glance it would seem that we should experience antialiasing since we are sampling at a rate that is less than Harry Nyquist's requirements of $f_s = 2 f_c$. This is not the case however, since the signal is band-limited first to remove any signals that would cause aliasing. What is important is the bandwidth of the passband signal, which is twice the bandwidth of the modulation signal, i.e., $2 \times 10\,\text{kHz} = 20\,\text{kHz}$. The sampling frequency range, in order to avoid alias overlapping and spectral inversion, is defined as

$$\frac{2 f_H}{m + 1} < f_s < \frac{2 f_L}{m}. \tag{6.1}$$

The upper side frequency is f_H and the lower side frequency is f_L, where m is a positive integer [ref: 7 Appendix A]. Constraints are placed on the value of m to avoid spectral inversion

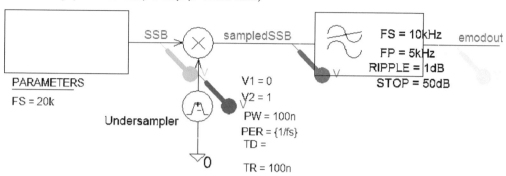

FIGURE 6.2: Down-sampling an SSB signal.

and aliasing, such as making m a positive even integer. However, we must always ensure that the sampling rate constraint is met, i.e.,

$$f_s \geq 2B \geq 2(f_H - f_L). \tag{6.2}$$

Here B is the signal bandwidth. Optimum sampling is achieved when we use an even positive value for m in (6.1).

6.2.2 Down-Sampling a Single-Sideband Signal

Fig. 6.2 shows the upper sidebands centered on a 1 MHz carrier defined in an **ABM** part. This is then sampled at 20 kHz and results in the two modulating signals appearing at baseband.

Set the **Analysis Setup/Transient** parameters as follows: **Run to time** = 10 ms, **Maximum step size** = 100 ns. Press **F11** to simulate and produce the signals shown in Fig. 6.3. From the **PROBE File** menu, tick **Run Commands** and select **Fig.6-003.cmd** to automatically produce the keystrokes necessary for the correct display.

The schematic in Fig. 6.4 is used to investigate the problems associated with down-converting an AM signal to baseband. The schematic comprises four sections: Double Sideband Full Carrier (DSBSC) production: down-converting with a pre-filter to remove the lower sideband, down-converting with a high-pass filter to remove the lower sideband, and the vector modulator, which is the best solution for removing image frequency signals. The vector demodulator consists of two **LOPASS** ABM filters (Passband frequency = 20 kHz), whose outputs are applied to a two-input **ABM2** part, where **EXP1** = sqrt(V(%IN1)*V(%IN1) and **EXP2** = +V(%IN2)*V(%IN2)) are added together producing an output $V_{\text{total}} = \sqrt{V_Q^2 + V_I^2}$.

The output spectra for each modulating system are shown in Fig. 6.5.

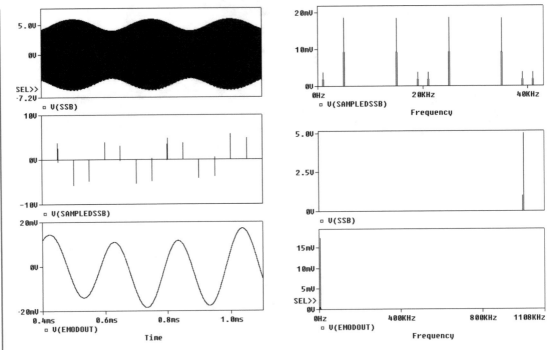

FIGURE 6.3: SSB and associated sampling waveforms.

6.3   DIGITAL RECEIVER

Fig. 6.6 shows the main parts of a digital receiver: RF stage, local oscillator, mixer and decimating filter, and demodulator. An AM signal is applied using an **ABM** part with the AM signal defined in the **EXP1** box. In this simple example, we make the carrier frequency quite low to speed up simulation times. The AM signal is sampled at 2.5 times the input carrier frequency. The local oscillator is a quadrature pair of oscillators, where the cosine and sine oscillator frequencies are set to the carrier frequency. The Q and I multiplier outputs are then FIR-filtered (cut-off frequency = 1 kHz), and a decimation stage reduces the number of samples by $M = 50$ kHz/1 kHz = 50. The quadrature outputs from each decimator are then applied to the demodulator to recover the baseband signal. The demodulator consists of two **LOPASS ABM** filters (the passband is set to 1 kHz), whose outputs are applied to a two-input **ABM2** part, where **EXP1** = sqrt(V(%IN1)*V(%IN1) is added to **EXP2** = + V(%IN2)*V(%IN2)). What this part is doing is simply producing an output as $V_{\text{total}} = \sqrt{V_Q^2 + V_I^2}$.

The AM signal (1 + 0.5*(sin(2*pi*fm*TIME)))*cos(2*pi*fc*TIME) includes a 1 kHz modulating signal and a 50 kHz carrier. In this example, we decimate by an M factor of 50, i.e., 50 kHz down to 1 kHz. The passband edge frequency is 1 kHz, and the decimation filter

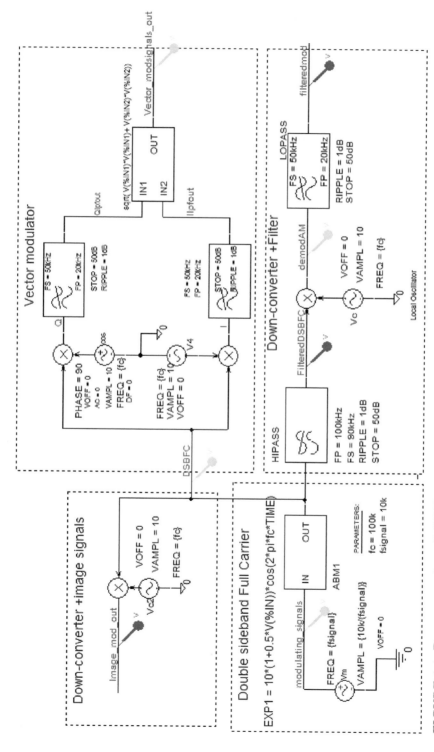

FIGURE 6.4: Down-converting a DSBFC signal

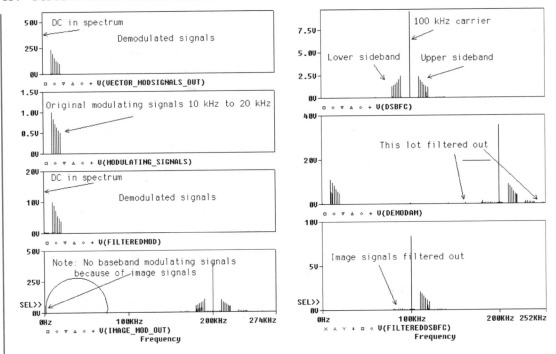

FIGURE 6.5: Down-converter spectra

stopband edge frequency is

$$f_s = F_{new\_output\_sample\_rate} - \frac{F_{original\_samplerate}}{2M} = 1\ kHz - \frac{50\ kHz}{2 \times 50} = 0.5\ kHz.$$

The decimation filter uses the **Laplace** part to define the delays and coefficients. A Matlab mfile obtain values for the filter tap values shown in the **PARAM** part. **Analysis Set up/Transient** parameters: **Output File Options/Print values in the output file** = 1 us, **Run to time** = 10 ms, **Maximum step size** = 0.1 us. Press **F11** to simulate and produce the signals shown in Fig. 6.7. From the **PROBE File** menu tick **Run Commands** and select **Fig.6-006.cmd** to automatically separate the plotted variables.

The time-domain signals are shown in Fig. 6.8.

It is left as an exercise to try down sampling in the first sampler. Set the sampler to the carrier frequency and see if the AM is shifted down to baseband.

6.4 DSP AND MUSIC

DSP is very useful for producing all sorts of musical effects such as reverb, echo, chorus, flanging, etc. Fig. 6.9 is a very simple echo unit based on an IIR filter design. Change the input wire segment to **speechout** and set the **Analysis Setup/Transient** parameters:

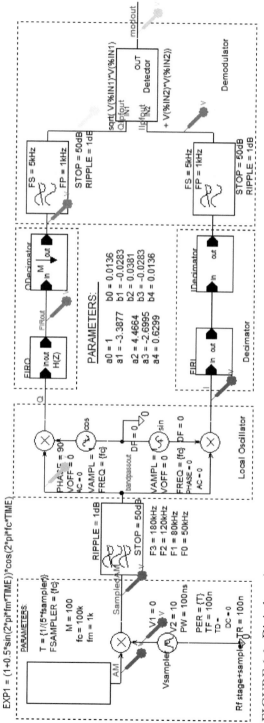

FIGURE 6.6: Digital receiver.

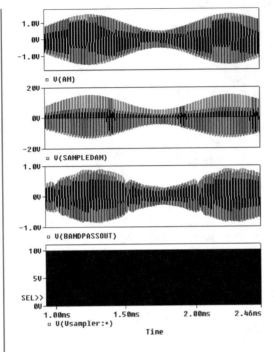

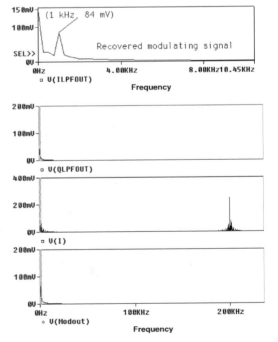

FIGURE 6.7: AM and down-sampled AM.

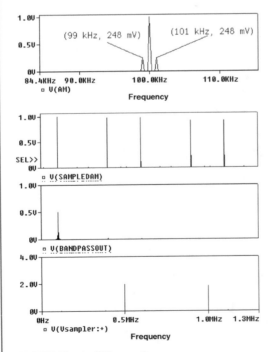

FIGURE 6.8: RF sampling spectra.

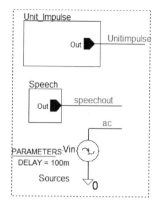
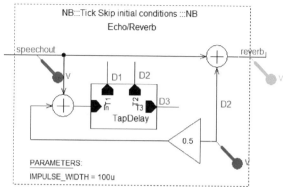

FIGURE 6.9: Echo/reverb.

Output File Options/Print values in the output file = 100 ns, **Run to time** = 3 s. Press **F11** to simulate.

The delayed signals are shown in Fig. 6.10.

Modify the schematic in Fig. 6.9 to produce the nine-tap delay unit shown in Fig. 6.11.

The inside of the delay block contains nine transmission lines as shown in Fig. 6.12. Note that the termination at both ends is equal to the characteristic impedance.

Set the Analysis tab to Analysis type: AC Sweep/Noise, AC Sweep Type to **Linear**, **Start Frequency** = 1, **End Frequency** = 1k, **Total Points** = 1000. Press **F11** to simulate and produce the comb-like frequency response shown in Fig. 6.13. This type of echo produces undesirable coloration in the signal due to the "teeth, or comb-like response" causing uneven frequent component distortion.

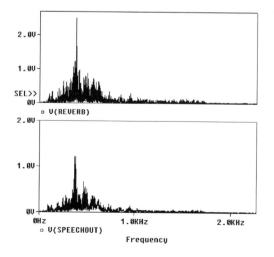
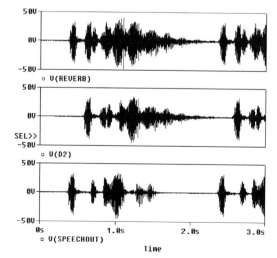

FIGURE 6.10: Reverberation signals.

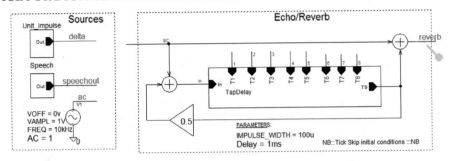

FIGURE 6.11: Multitap delay echo unit.

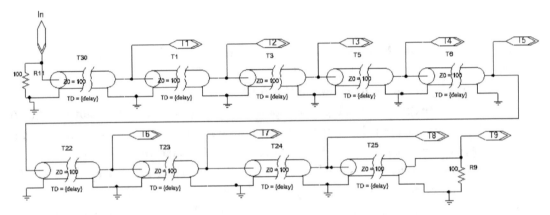

FIGURE 6.12: Nine-delay line.

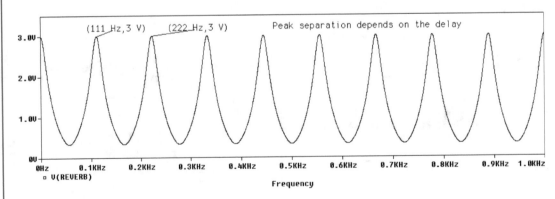

FIGURE 6.13: Comb frequency response.

6.4.1 Phasing Effect

A useful musical effect called phasing produces a "whooshing sound," and is achieved using a notch filter (a band-stop filter) employed in the feedback loop shown in Fig. 6.14.

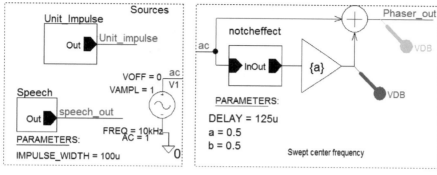

FIGURE 6.14: Phasing effect.

The schematic inside the **notcheffect** block is shown in Fig. 6.15.
The notch filter transfer function is

$$H(z) = \frac{Y(z)}{X(z)} = \left(\frac{1+a}{2}\right)\left(\frac{1 - 2bz^{-1} + z^{-2}}{1 - b(1+a)z^{-1} + az^{-2}}\right). \tag{6.3}$$

The center frequency, f_0, is expressed as

$$b = \cos 2\pi f_o \Rightarrow f_o = \frac{1}{2\pi}\cos^{-1} b. \tag{6.4}$$

The Q-factor (notch tightness) is calculated as

$$Q = \frac{f_0}{\Delta f} = \frac{\cos^{-1} b}{\cos^{-1}[2a/(1+a^2)]}. \tag{6.5}$$

Calculate the resonant frequency, bandwidth and Q-factor for $a = b = 0.5$, and compare to the simulated results. Change the input wire segment name to **ac,** and carry out an AC analysis: **Analysis Setup, AC Sweep/Linear, Points/Decade** = 10001, **Start Frequency** = 10 Hz, and **End Frequency** = 8 kHz. Test the notch filter by applying an AC signal to the circuit in Fig. 6.15, and a dB marker on the output. The filter coefficients are defined in the **PARAM** part. Ticking the **Parametric** box in the **Analysis Setup** enables us to vary the a coefficient

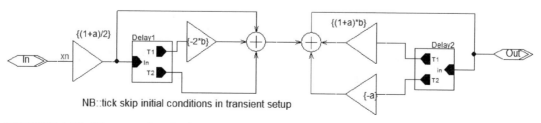

FIGURE 6.15: The second-order bandstop filter.

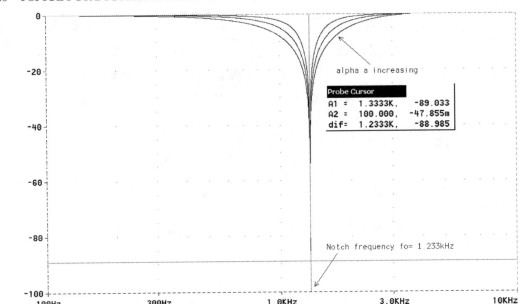

FIGURE 6.16: Varying α.

first from 0.1 to 0.9, in steps of 0.1. Simulate by pressing the **F11** key to produce the frequency response shown in Fig. 6.16.

The resonant, or notch frequency, is changed by varying b for a range of values, as illustrated in Fig. 6.17. Is the Q-factor changing?

6.5   ZERO-FORCING EQUALIZER

A receiver equalizer eliminates intersymbol interference (ISI is the interference between received pulses) when the equalizer response is the inverse of the actual channel response. To demonstrate this, consider a channel with a transfer function $H(z)$ and an equalizer transfer function $L_{EQ}(z)$ that is the reciprocal of the channel transfer function:

$$L_{EQ}(z) = \frac{1}{H(z)}. \tag{6.6}$$

Fig. 6.18 shows how to model a multipath interference signal by adding the delayed speech signal (an echo) to the original signal. The filter transfer function is

$$H(z) = \frac{Y(z)}{X(z)} = 1 + 0.75z^{-1}. \tag{6.7}$$

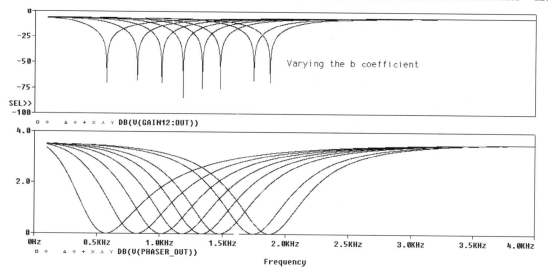

FIGURE 6.17: Varying the b coefficient.

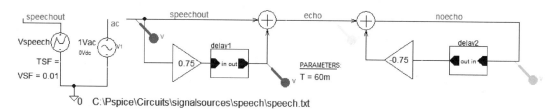

FIGURE 6.18: Echo canceler.

Export the signal and listen to the echo signal. The "equalizer filter" transfer function is

$$H'(z) = \frac{1}{H(z)} = \frac{1}{1 + 0.75z^{-1}}. \tag{6.8}$$

Set the Analysis tab to Analysis type: **Time Domain** (Transient), **Run to time** $= 10$ s and press **F11** to produce Fig. 6.19. Note the echo has now been cancelled.

The complete transfer function for the "equalized path" has a value of 1, i.e.,

$$H(z) = \frac{1}{(1 + 0.75z^{-1})}(1 + 0.75z^{-1}) = 1. \tag{6.9}$$

Export the signal, and verify that the echo has been canceled by listening to the output "noecho" signal using the WAV2ASCII program. Set up a transient analysis and plot the signals. To obtain the system impulse response for a real system, we transmit a signal and compute the inverse filter from the response.

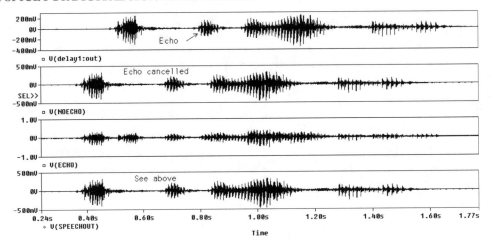

FIGURE 6.19: Echo canceler.

6.5.1   Three-Tap Zero-Forcing Equalizer

The three-tap zero-forcing equalizer in Fig. 6.20 attempts to eliminate ISI by minimizing peak distortion using *certain tap coefficients*. The coefficients make the equalizer output zero on either side of the desired pulse but equal to the input pulse at the desired location. The equalizer uses a transversal filter consisting of a series of delays at T s intervals, where T is the symbol width. The summed output of the equalizer is sampled at the symbol rate and fed to a decision device. The equalizer output is expressed in terms of the input $v_{in}(t)$ and tap coefficients, c_n, as

$$V_{out}(t) = \sum_{k=0}^{2N} c_k V_{in}(t - kT). \tag{6.10}$$

The two-symbol delay in the ZFE uses two transmission lines suitably terminated as shown in Fig. 6.21.

The input data signal in Fig. 6.22 is applied using a **VPWL_FILE** part, where the text file is located at C:\Pspice\Circuits\signalsources\data\equaliser2.txt.

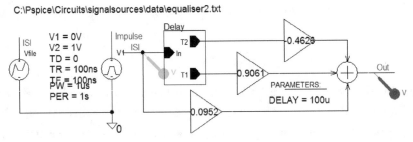

FIGURE 6.20: Transversal FIR-type equalizer.

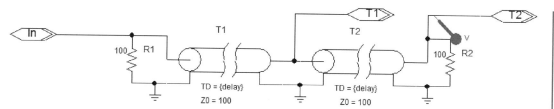

FIGURE 6.21: The two-delay elements using transmission lines.

equaliser.txt - MicroSim Text Editor

File   Edit   Search   View   Insert   Help

0s	0.01
100us	0.01
100.1us	-0.1
200us	-0.1
200.1us	1
300us	1
300.1us	0.5
400us	0.5
400.1us	0.1
500us	0.1
500.1us	0.0

FIGURE 6.22: The input ISI signal.

6.5.2   Example

The input pulse train is of length $(2N - 1) = (2 \times 3 - 1) = 5$. For successive time intervals, (t) to $(t - T)$ is 0.01, −0.1, 1, 0.5, 0.1. Calculate the equalizer coefficients C_0, C_1, and C_2 (some books refer to these coefficients as C_{-1}, C_0, C_1), to optimize the linear three-tap FIR transversal feed-forward equalizer such that the output is 0 1 0.

6.5.3   Solution

The desirable output at $n = 0, 1, 2$, is 0, 1, 0. More coefficients are required if we wish to force other input values to zero. The set of equations are

$$\begin{bmatrix} 0.01 & 0.0 & 0.0 & 0.0 & 0.0 \\ -0.1 & 0.01 & 0.0 & 0.0 & 0.0 \\ 1.0 & -0.1 & 0.01 & 0.0 & 0.0 \\ 0.5 & 1.0 & -0.1 & 0.01 & 0.0 \\ 0.1 & 0.5 & 1.0 & -0.1 & 0.01 \\ 0.0 & 0.1 & 0.5 & 1.0 & -0.1 \\ 0.0 & 0.0 & 0.1 & 0.5 & 1.0 \\ 0.0 & 0.0 & 0.0 & 0.1 & 0.5 \\ 0.0 & 0.0 & 0.0 & 0.0 & 0.1 \end{bmatrix} \cdot \qquad (6.11)$$

We need to reduce this matrix by eliminating the top and bottom N rows and turning it into a square matrix of size $(2N+1)$ by $(2N+1)$ to the matrix outlined in bold to yield a matrix of three linear simultaneous equations in three unknowns:

$$
\begin{array}{c} v_{out}(t) = \\ v_{out}(t-T) = \\ v_{out}(t-2T) = \end{array}
\begin{bmatrix} 0 \\ 1 \\ 0 \end{bmatrix} =
\begin{array}{ccc} 1.0C_0 & -0.1C_1 & +0.01C_2 \\ 0.5C_0 & +1.0C_1 & -0.1C_2 \\ 0.1C_0 & +0.5C_1 & +1.0C_2 \end{array} =
\begin{bmatrix} 1.0 & -0.1 & 0.01 \\ 0.5 & 1.0 & -0.1 \\ 0.1 & 0.5 & 1.0 \end{bmatrix}
\begin{bmatrix} C_0 \\ C_1 \\ C_2 \end{bmatrix}
\tag{6.12}
$$

$$
C_0 = \frac{\begin{bmatrix} 0 & -0.1 & 0.01 \\ 1 & 1.0 & -0.1 \\ 0 & 0.5 & 1.0 \end{bmatrix}}{\begin{bmatrix} 1.0 & -0.1 & 0.01 \\ 0.5 & 1.0 & -0.1 \\ 0.1 & 0.5 & 1.0 \end{bmatrix}}
$$

$$
= \frac{(-1)[(-0.1)(1.0)-(0.01)(0.5)]}{1\begin{bmatrix} 1 & -0.1 \\ 0.5 & 1 \end{bmatrix} - (-0.1)\begin{bmatrix} 0.5 & -0.1 \\ 0.1 & 1 \end{bmatrix} + 0.01\begin{bmatrix} 0.5 & 1 \\ 0.1 & 0.5 \end{bmatrix}}
\tag{6.13}
$$

$$
C_0 = \frac{(-1)[(-0.1)(1.0)-(-0.01)(0.5)]}{1(1+0.05)+0.1(0.5+0.01)+0.01(0.25-0.1)} = \frac{0.105}{1.1025} = 0.0952.
\tag{6.14}
$$

These equations are solved using Matlab as $\underline{A}\underline{C} = \underline{B}$, with the solution $\underline{C} = A^{-1}\underline{B}$. Enter the following matrices into Matlab. The command lines are given below:

≫ sigdistort =[
1 −0.1 0.01
0.5 1 −0.1
0.1 0.5 1].

The solution to the simultaneous equations is obtained by multiplying the coefficient matrix inverse by the data matrix. The inverse matrix uses the Matlab command inv (), e.g., ≫ inv(sigdistort)

ans =
  0.9524 0.0952 0
  −0.4626 0.9061 0.0952
  0.1361 −0.4626 0.9524.

The desired data at the first three sample periods is defined as

$$\gg data = [0$$
$$1$$
$$0].$$

The coefficients are then calculated using

$\gg inv(sigdistort)*data$

ans =

0.0952

0.9061

−0.4626

$C_0 = 0.0952$, $C_1 = 0.9061$ and $C_2 = -0.4626$.

Repeat this procedure for the other coefficients. These values produce a pulse somewhat closer to the original pulse. With only three coefficients, we cannot predict the shape of the pulse at sample times, other than at the pulse center, and one interval on either side. This equalizer is trained by sending a training signal at the start of a transmission enabling the coefficients to be calculated. These values remain fixed until a break in transmission set the **Transient Analysis** parameters as follows: **Run to time** = 1 ms, **Maximum step size** = 100 ns. Press **F11** to simulate. Compare the input and output signals in Fig. 6.23 to show how ZFE reduces ISI.

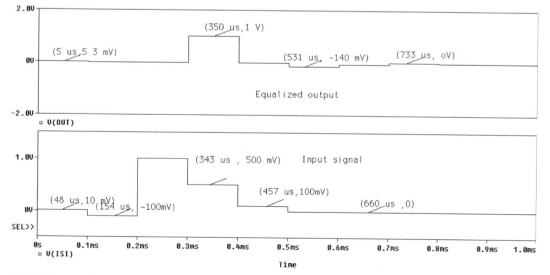

FIGURE 6.23: The improved output signal.

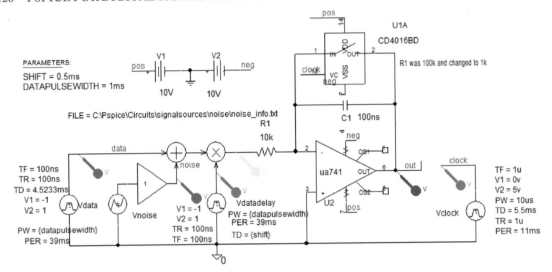

FIGURE 6.24: Analog correlator.

6.6 EXERCISES

1. Fig. 6.24 shows an "integrate and dump" correlator circuit. The data is corrupted by noise and multiplied by the same data but shifted by an amount "shift" using the **PARAM** part. Note the use of the **IC** part for setting the initial conditions. The NRZ-B data has each zero -1, and each binary one $+1$, and yields a unity autocorrelation function (AC) for zero delay, and $-1/(2^m - 1)$ for a delay greater that one bit. The AC function is triangular shaped between the limits -1 bit and $+1$ bit, and is centered at zero. N is the sequence length, and d is the shifting parameter, where $0 \leq d \leq N-1$. The correlator waveforms are displayed in Fig. 6.25.
   The output of the correlator rises as the swept pulse overlaps with the original pulse buried in the noise.

2. Test the run properties of the PRBS generator using the circuits shown in Figs. 6.26 and 6.27.

3. Investigate the seven-tap zero-forcing equalizer shown in Fig. 6.28. A larger number of taps are required in order to improve on the three-tap design. The input signal length is $2N - 1 = 13$ defined as $0, 0, 0.01, -0.01, 1, 0.5, 0.1, -0.1, 0.5, 0.1, 0, 0, 0$. The desired signal is $0, 0, 0, 1, 0, 0, 0$.
   The impulse response is shown in Fig. 6.29.
   The good signals are shown in Fig. 6.30.

4. Adaptive equalizers minimize ISI under dynamic conditions and automatically update their parameters during data transmission. A training data sequence is transmitted and

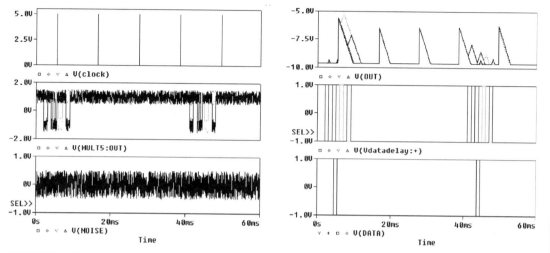

FIGURE 6.25: Analog correlator signals.

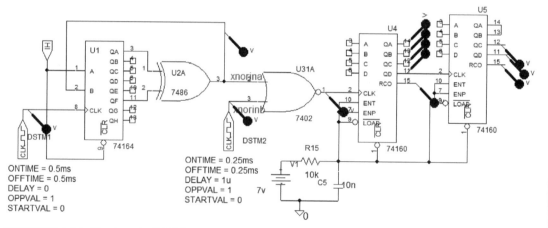

FIGURE 6.26: Testing the PRBS properties (a).

is compared to a pseudorandom bit sequence (PRBS) in the receiver and produces an error voltage to optimize the coefficients for minimum distortion. Between training periods the filter coefficients are not updated. This is a limitation because the channel characteristics could change during the period when the coefficients are being updated. There are many algorithms for adapting the FIR filter coefficients, of which the most widely used one is the least mean squares (LMS). Here an adaptive, or blind algorithm, operates on the signal and requires no training sequence.

Fig. 6.31 is an adaptive equalizer, with an input training signal, $x(n)$, applied to the unknown system, which is modeled as an FIR filter with adjustable coefficients. The FIR filter system has known coefficients. The outputs from both filters are subtracted

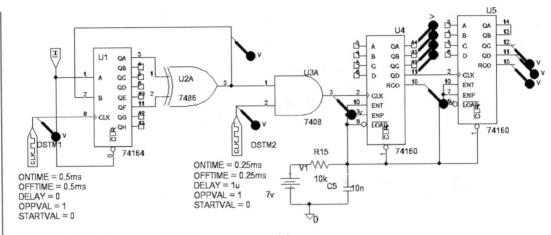

FIGURE 6.27: Testing the PRBS run properties (b).

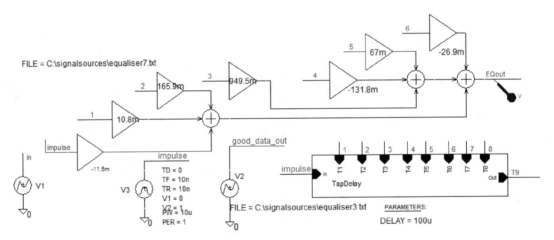

FIGURE 6.28: Seven-tap transverse FIR filter.

to generate an error signal, $e(n)$, that is fed back to the adaptive algorithm input and changes the filter coefficients until the error signal is sufficiently small. The resultant FIR filter response now represents that of the previously unknown system. On channels whose frequency-response characteristics are unknown, but time invariant, we measure the channel characteristics and adjust the equalizer parameters. Once adjusted, the parameters remain fixed for the duration of the data transmission.

The inside of the FIR filter block is shown in Fig. 6.32.

The unknown FIR filter is shown in Fig. 6.33.

The filter weights are updated according to the algorithm,

$$w1new = w1new + c.error.x_{in}, \qquad (6.15)$$

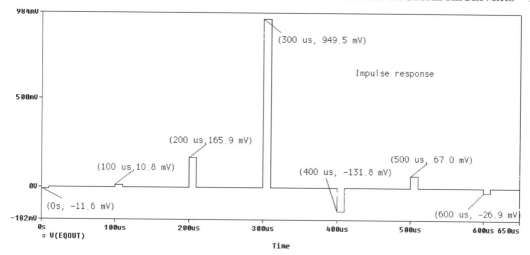

FIGURE 6.29: The impulse response.

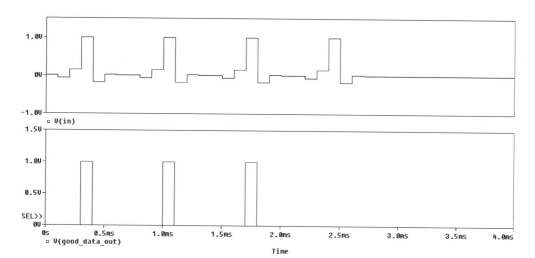

FIGURE 6.30: The good signals

where c is the learning constant and *error* is the difference between the two filter outputs. The weight coefficients for one filter are known, but the other filter coefficients are unknown but updated according to (6.15). The **ABM** part in Fig. 6.34 updates the filter coefficients using an IF THEN ELSE statement as

$$If(V(\%IN) <= V(w2new), (V(w2new)$$
$$-V(deltaweight2)), (V(w2new) + V(deltaweight2))). \qquad (6.16)$$

The adaptive response to a DC input is shown in Fig. 6.35.

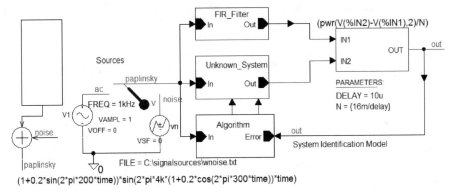

FIGURE 6.31: Adaptive equalizer for system identification.

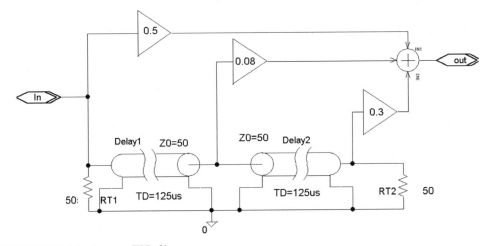

FIGURE 6.32: The known FIR filter.

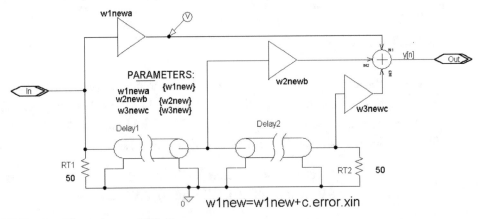

FIGURE 6.33: The unknown FIR filter.

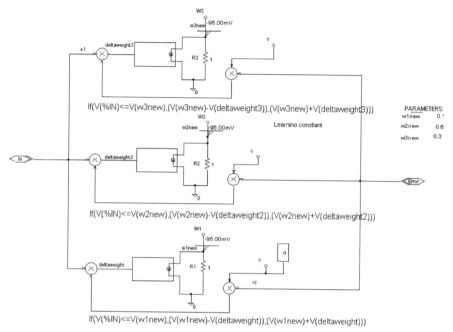

FIGURE 6.34: The adaptive algorithm.

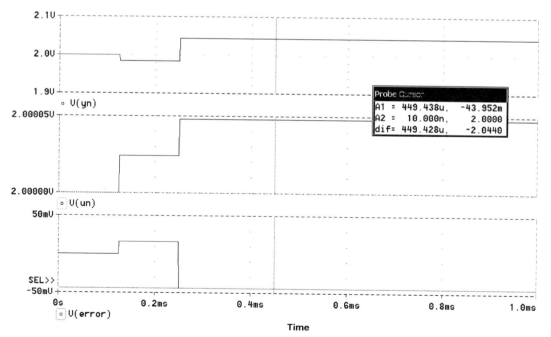

FIGURE 6.35: The adaptive signals for a constant DC input.

Appendix A: References

BOOKS

1. Tobin P (Mar 2007) PSpice for Filters and Transmission Lines. Morgan Claypool publishers.

2. Tobin P (Mar 2007) PSpice for Digital Communications Engineering. Morgan Claypool publishers.

3. Bateman A and Yates W (1989) Digital Signal Processing Design. Computer Science Press.

4. Young PH (1991) Electronic Communications Techniques. USA: Maxwell-MacMillan.

5. Digital Signal Processing: A Practical Approach, Second Edition Emmanuel C Ifeachor and Barrie W Jervis ISBN 0 201 59619 9.

6. Concepts In Systems and Signals by John D Sherrick.

7. Tobin Paul PSpice for Circuit theory and Electric devices Morgan Claypool publishers, Jan 2007.

8. Tobin Paul PSpice for Analog Communications Engineering Morgan Claypool publishers, Mar 2007.

9. Sanjit K Mitra Digital Signal Processing: A Computer-Based Approach, 2e with DSP.

INTERNET

- http://en.wikipedia.org/wiki/Shannon-Hartley_theorem
- http://www.dspguru.com/info/faqs/multrate/resamp.htm
- http://www.analog.com/processors/resources/beginnersGuide/index.html
- http://www.arrl.org/tis/info/sdr.html
- http://www.freqdev.com/guide/dspguide.html#digdesign
- http://www.bores.com/courses/intro/program/index.htm
- http://www.dsprelated.com/showmessage/20912/1.php
- http://web.mit.edu/6.555/www/fir.html
- http://www.analog-innovations.com

Appendix B: Tables

TABLE B-1: Useful Trigonometric Identities

FUNCTION	EXPANSION
$\sin A \sin B$	$0.5\cos(A-B) - 0.5\cos(A+B)$
$\cos A \cos B$	$0.5\cos(A-B) + 0.5\cos(A+B)$
$\sin A \cos B$	$0.5\sin(A-B) + 0.5\sin(A+B)$
$e^{\pm j\theta}$	$\cos\theta \pm j\sin\theta$
$\cos\theta$	$(e^{j\theta} + e^{-j\theta})/2$
$\sin\theta$	$(e^{j\theta} - e^{-j\theta})/2j$

$$1 = \cos^2\theta + \sin^2\theta$$

$$\cos 2\theta = \cos^2\theta - \sin^2\theta$$

TABLE B-2: Laplace and z-Transform Table

FUNCTION	$f(t)$	LAPLACE TRANSFORM	$f(n)$	Z-TRANSFORM ($t = nT = n$)
Unit step	$u(t)$	$1/s$	$u(n)$	$z/z-1$
Unit impulse	$\delta(t)$	1	$\delta(n)$	1
Unit ramp	T	$1/s^2$	N	$nz/(z-1)^2$
Polynomial	t^n	$n!/s^{n+1}$	t^n	$T^2 z(z+1)/(z-1)^2$ where $n=2$
Decaying exponential	e^{-at}	$1/(s+a)$	$e^{-an}u(n)$	$z/z-e^{-an}$
Growing exponential	$1/a(1-e^{-at})$	$1/(s+a)(s)$	$1/a(1-e^{-an})$	$z(1-e^{-an})/a(z-1)(z-e^{-an})$
Sine	$\sin(\omega t)$	$\omega/(s^2+\omega^2)$	$\sin(n\theta)u(n)$	$\dfrac{z\sin n\theta}{z^2 - 2z\sin n\theta + 1}$
Cosine	$\cos(\omega t)$	$s/(s^2+\omega^2)$	$\cos(n\theta)u(n)$	$\dfrac{z(z-\cos n\theta)}{z^2 - 2z\cos n\theta + 1}$
Damped sine	$e^{-at}\sin(\omega t)$	$\omega/[(s+a)^2+\omega^2]$	$e^{-an}\sin(n\theta)$	$\dfrac{ze^{-an}\sin(n\theta)}{z^2 - 2ze^{-an}\cos n\theta + e^{-2an}}$
Damped cosine	$e^{-at}\cos(\omega t)$	$(s+a)/[(s+a)^2+\omega^2]$	$e^{-an}\cos(n\theta)$	$\dfrac{z^2 - ze^{-an}\cos(n\theta)}{z^2 - 2ze^{-an}\cos n\theta + e^{-2an}}$
Delay	$f(t-k)$	e^{-sk}	$f(n-k)$	z^{-k}

Index

6 dB rule, 2

adaptive equaliser, 126
append waveform, 9
autocorrelation function, 126

bandpass digital filter, 42
bandpass filter, 31, 42
Bartlett, 76
bibo stable, 17
bilinear transform, 61, 86
blind algorithm, 127

causal, 17
characteristic impedance, 7
Chebychev, 72
chorus, 114
cir netlist, 38
comb response, 117
correlation, 51
cut-off frequency, 23

dac, 1
decimation, 96
decimator, 103
delay, 37
DENOM, 39
denom, 38
difference equation, 17, 19
digital convolution, 43
digital elliptic filter, 60

digital frequency, 3
digital oscillator, 48
digital receivers, 109
digital state variable filter, 51
digital State variable Quad oscillator, 51
display evaluation, 75

Edit Properties, 39
elliptical filter, 35
equi-ripple digital filter, 57
Euler's identity, 25
evaluation measurement, 75

feed forward, 24
filter coefficients, 13, 18
filtercoef, 23
finite-impulse response, 18, 53
fir, 17
fircos.m, 85
flanging, 114
flip and slip, 43
Fourier method, 75
func statement', 36

gain part, 23
Gibbs phenomenon, 70, 72, 81
global parameter, 23
group delay, 26, 40

Hamming window, 72, 82
Hann window, 72, 82

hierarchical blocks, 5
Hilbert transformer, 90

ic part, 126
if, then else, 72, 76, 129
iir, 17
impulse invariance digital filter,
    65
impulse invariant mapping, 65, 70
impulse response, 29
Impulse_width, 11
infinite impulse response, 18, 53
integrate and dump correlator circuit,
    126
integrator, 95
interpolation, 96, 102

L'Hopital's rule, 83
Laplace part, 24, 37, 85
Least Mean Squares (LMS),
    127
linear, 17

maximum step size, 41
multi- rate sampling, 95

notch filter, 118
null, 11

one-sided z-transform, 14, 20

param part, 12, 23, 26, 38
parametric, 23
passband gain, 25
phase shift method, 93
pins icon, 6

pole-zero map, 49
pre-warp, 60
PSpicead.exe., 36

Q-factor, 34
quadrature carriers, 89
quantisation noise, 2

recursive filter, 18, 24, 53
recursive system, 29
reverberation, 114
ripple magnitude, 81
roots.m, 31
run commands, 46, 93, 99

sampling, 37
sampling period, 14, 20, 103
sampling theorem, 7
set up block, 10
sinc-shaped, 4
spectral leakage, 75
SQNR, 2
ssbsc, 93
step response, 28
stopband edge frequency, 98
storage requirements, 99, 101
system identification filter,
    127

time-invariant, 17
trace goal function, 33
transversal FIR filter, 83
Types I and II, 54

unit impulse, 14
unit impulse function, 9

unit step function, 9, 14
unsynchronize, 50
upsampling, 102

vpulse generator, 9
vpulse part, 28

windowing, 72
wire segment name, 18

zero-forcing equaliser, 122
ZFE, 125
zplane.m, 22, 50

Author Biographies

Mr. Paul Tobin graduated from Kevin Street College of Technology (now the Dublin Institute of Technology) with honours in electronic engineering and went to work for the Irish National Telecommunications company. Here, he was involved in redesigning the analogue junction network replacing cables with PCM systems over optical fibres. He gave a paper on the design of this new digital junction network to the Institute of Engineers of Ireland in 1982 and was awarded a Smith testimonial for one of the best papers that year. Having taught part-time courses in telecommunications systems in Kevin Street, he was invited to apply for a full-time lecture post. He accepted and started lecturing full time in 1983. Over the last twenty years he has given courses in telecommunications, digital signal processing and circuit theory.

He graduated with honours in 1998 having completed a taught MSc in various DSP topics and a project using the Wavelet Transform and neural networks to classify EEG (brain waves) associated with different mental tasks. He has been a 'guest professor' in the Institut Universitaire de Technologie (IUT), Bethune, France for the past four years giving courses in PSpice simulation topics. He wrote an unpublished book on PSpice but was persuaded by Joel Claypool (of Morgan and Claypool Publishers) at an engineering conference in Puerto Rico (July 2006), to break it into five PSpice books (One of the books introduces a novel way of teaching DSP using PSpice). There are over 500 worked examples in the five books covering a range of topics with theory and simulation results from basic circuit theory right up to advanced communication principles. Most of the worked example circuits have been thoroughly 'student tested' by Irish and International students and should mean little or no errors but alas. . .

Printed in the United States
by Baker & Taylor Publisher Services